T0215326

"In the context of the multiple and conjoined crises, contradictions, contingencies, and extremities facing our planet, *Sustainable Places* is an original and important statement on the power of place to act as that fulcrum of hope to shape a better foundational world. Rich in conceptual framings and empirical nuances, this book is recommended reading for boffins, bureaucrats and all parties interested in understanding why and where geography increasingly matters. Carbon-based neoliberal global capitalism is in crisis, but there are alternatives".

Professor Martin Jones,
Vice-Chancellor, Staffordshire University

"The Deep Place approach offers hope at a moment when the decades-long push to globalise everything is crumbling under the pressures of war, disease, and social tensions of private wealth and social impoverishment. Having long pioneered theory and praxis of sustainable place-making, the authors help us shift away from a negative, fearful mindset – one that fails to see the possibility to change the institutions that inflict harms on one another and on earthly conditions of all life. Beginning with each specific place, human groups can reconnect in life-enhancing ways both within places of all sizes, and across places. This work offers inspiring examples of embryonic practices to guide democratic transitions towards collective wealth and wellbeing".

Professor Harriet Friedmann,
University of Toronto

SUSTAINABLE PLACES

At the intersection of environmental sustainability, economic and social disintegration, and regeneration, this book offers a new engaged methodology and approach that problematises spatial and social inequality but also offers a way forward for local communities as the testbed for sustainability.

This book calls for more holistic place-based action to address the social and environmental crisis, deploying the Deep Place approach as one contribution to the toolbox of actions that will underpin the UN Decade of Action towards the Sustainable Development Goals. The authors suggest that 'place' is a critical window on how to conceive a resolution to the multiple and overlapping crises. As well as diagnosing the problem (the world as it is), this book also offers a normative advocacy (the world as it could/should be and proposed pathways to get there). A series of 'Deep Place' case studies from the UK, Australia, and Vanuatu help to illustrate this approach. Ultimately, the book argues for the need for a real and green 'new deal' and identifies what this should be like. It suggests that a new economic order, whilst eventually inevitable, requires radical change. This will not be easy but will be essential given the current impasse, caused, not least by the conjunction of carbon-based, neoliberal capitalism in crisis and the multifactorial global ecological crisis. Ultimately, it concludes that there is a need to develop a new model of 'regenerative collectivism' to overcome these crises.

This book will be of interest to academics, policy practitioners, and social and climate justice advocates/activists.

David Adamson is an Honorary Professor at the College of Health, University of Newcastle, Australia, and Emeritus Professor at the University of South Wales, UK.

Lorena Axinte is a Research Associate at the School of Geography and Planning, Cardiff University, UK.

Mark Lang is an Honorary University Associate at the School of Law and Politics, Cardiff University, UK.

Terry Marsden is an Emeritus Professor of Environmental Policy and Planning at the School of Geography and Planning, Cardiff University, UK.

Routledge Explorations in Environmental Studies

For more information about this series, please visit: www.routledge.com/Routledge-Explorations-in-Environmental-Studies/book-series/REES

SUSTAINABLE PLACES

Addressing Social Inequality
and Environmental Crisis

David Adamson

Lorena Axinte

Mark Lang

Terry Marsden

Routledge
Taylor & Francis Group

LONDON AND NEW YORK

from Routledge

Cover image: Misha Kaminsky

First published 2023
by Routledge
4 Park Square, Milton Park, Abingdon, Oxon OX14 4RN

and by Routledge
605 Third Avenue, New York, NY 10158

Routledge is an imprint of the Taylor & Francis Group, an informa business

British Library Cataloguing-in-Publication Data
A catalogue record for this book is available from the British Library

Library of Congress Cataloging-in-Publication Data
A catalog record has been requested for this book

ISBN: 978-1-032-11794-2 (hbk)
ISBN: 978-1-032-11791-1 (pbk)
ISBN: 978-1-003-22155-5 (ebk)

DOI: 10.4324/9781003221555

Typeset in Bembo
by codeMantra

CONTENTS

ACKNOWLEDGEMENTS

The research informing this book and the various Deep Place studies outlined in Part Two have been developed by the authors over many years. We would like to thank the communities, organisations, and colleagues who have supported and engaged with us on this journey. We would particularly like to thank the Centre for Regeneration Excellence Wales, Cardiff University, Caerphilly County Borough Council, and Compass Housing, without whom it would have been difficult to undertake our Deep Place studies.

ACKNOWLEDGMENTS

PART ONE

1

GLOBAL CRISIS

Our moment of reckoning

Every one of us has our origins in a physical space, a location where we begin our human development and establish relationships with family, friends, community, and our physical and natural environment. Depending on that location, we will play and mature in urban streets, municipal parks, rural woodland, or those fascinating combinations at the edge of both city and rurality. For those of us living beyond the boundaries of the so-called 'developed world', this life journey may engage with a transition from rural and traditional life patterns to the deprivations of the informal settlements of rapidly urbanising cities. In that dichotomy, we see laid bare the way that *place* conditions the very pattern of life itself and the opportunities and challenges that an individual may experience.

The influence exerted by place is, perhaps, less clear in the social structure of 'advanced' capitalist societies where the myth of social mobility, epitomised in the American Dream, suggests that those from the most humble and impoverished origins can reach the highest echelons of society. In reality, the place in which we are born and the social patterns dominant in that place will fundamentally shape many aspects of our lives (Rustin and Chamberlayne, 2002). Although not an iron cage from which there is no escape, our point of origin will exert huge influence over everything from our birthweight, our early social development, our educational outcomes, and the occupations we will follow. These, in turn, shape our incomes, the type of house we will occupy and the social circles we will move in. It will shape our patterns of consumption and production and shape our carbon footprint. Fundamentally, the places we occupy will determine our health (Marmot and Bell, 2012), the extent of our 'well-lives' in which we are able to enjoy all aspects of an active life, and most critically, our life expectancy, the ultimate 'quantity' of life we will experience.

DOI: 10.4324/9781003221555-2

This book is about how place can act as a powerful lens to understand and address the multiple and interconnected challenges faced by people and planet today. It comes at a time when there is a need for urgent action on climate change, biodiversity loss, and growing environmental vulnerabilities, if we are to avoid the extreme consequences identified in the 2021 Sixth Report of the Intergovernmental Panel on Climate Change (IPCC). The multiple impacts of the environmental crisis are geographically and spatially differentiated with significant vulnerabilities facing global regions, such as the Pacific. Within nations, particular places will experience diverse impacts of the climate crisis, from desertification to coastal inundation. Where we live has, perhaps, never been so important for determining our futures.

This book is also about the social crisis that runs in parallel and in close connection with the environmental crisis. This is the crisis of global poverty and inequality that exists both within and between nations. Again, the place offers a critical lens with which to examine the complex social and economic relationships that structure and mediate the distribution of resources and wealth at the planetary and local levels. Poverty characterises the lives of humans in the most advanced societies and in nations struggling to support their populations to achieve the most basic level of subsistence. Whilst these are different degrees of inequality, its impacts on those experiencing its effects are both corrosive and destructive to human well-being.

We also present a view that there are common causes of these conjoined crises. These can be found in a dominant social, economic, and political paradigm that has become a received wisdom and presents itself as the only way to organise human production, consumption, and trade. We identify neoliberalism and associated political models as a core causal element in both the climate crisis and the social crisis. Since the 1980s, states have increasingly supported a globalised system of trade, favoured the conditions for global capital accumulation and flow, and maximised the returns to the transnational corporations that have come to dominate global economic activity. This has been at the expense of the environment, as the economy and the pursuit of unrestricted 'growth' have moved beyond the planetary resource boundary and depleted and degraded the natural environment. Environmental protection is seen as a cost of production and a brake on growth. Additionally, to fund the favourable tax regimes and financial incentives to attract global capital, we have seen a diminution of national welfare rights and social protection systems in the welfare economies of the northern hemisphere, and rapid urbanisation processes in the nations emerging into these economic structures in the global south. Both have brought rising poverty levels across traditional economic divisions of 'developed' and 'undeveloped' worlds.

The inter-relationships between the environmental crisis and social crisis have been most clearly evidenced as the COVID-19 global pandemic has traversed the globe. The pandemic has acted as an alarm clock that has awoken the world to the frailty of the increasingly connected but unequal human civilisation. Although the urgency of addressing the immediate crisis has been focused on public health, it also has reminded us, or at least it should have reminded us, that our very existence is intimately connected to nature and environment. Over a century of deeply ingrained bias has led to a deluded human perception of our own invulnerability to environmental forces, which had previously only been marginally threatened by the occasional 'natural disaster'. COVID-19 has shaken this belief to its core. It has also revealed the deep social divisions and the under-lying patterns of poor health that are a central feature of poverty and inequality.

In 2021, the World Health Organisation (WHO) established a Scientific Advisory Group for the Origins of Novel Pathogens, and at its first meeting the WHO Director-General, Tedros Adhanom Ghebreyesus, clearly stated that 'outbreaks of viruses, known and unknown, are a fact of nature'. Moreover, a US investigation into the origins of COVID-19 in 2021, although proving in-conclusive, determined that the virus was not manmade (Maxmen, 2021). Our very destructive relationship with nature may, then, prove to be critical in un-derstanding the origins of new infectious diseases. Indeed, 'emerging infectious diseases frequently originate from pathogen spill-overs from wildlife to humans; contributing factors include forest fragmentation, habitat destruction, agricul-tural expansion, concentrated livestock production and human penetration into wildlife habitats' (Rulli et al., 2021). The intensifying urbanisation evident in 'developing' nations increases these risks, as the interface between human settle-ments and natural environments expands exponentially.

Although the precise origins of the COVID-19 coronavirus are likely to re-main contested for the foreseeable future, what is clear is that the human re-lationship with nature is now likely to be understood as a far more symbiotic one than has been previously assumed by many people. The UN Environment Programme Global Biodiversity Outlook 5 concluded that 'the COVID-19 pan-demic…reminds us all of the profound consequences to our own well-being and survival that can result from continued biodiversity loss and the degradation of ecosystems' (p. 8). Moreover, '…when we destroy and degrade biodiversity, we undermine the web of life and increase the risk of disease spill-over from wildlife to people' (Secretariat of the Convention on Biological Diversity, 2020, p. 24). The myths of onward 'progress' and, perhaps, ever-continuing growth have been spectacularly and unceremoniously busted.

The pandemic has also exacerbated and amplified the inequalities of wealth and income globally. Whilst many of us have faced the personal, emotional, and mental health impacts of multiple and prolonged 'lockdowns', isolation and related measures have done much at times to limit the transmission of COVID-19. Our capacity to endure such measures is, however, a function, in part, of our positional income, wealth, and housing quality. For those in low-income coun-tries, the necessity to earn an often subsistence level income renders self-isolation impossible (Bargain and Aminjonov, 2020). Housing quality, sanitation condi-tions, and overcrowding render the social distancing and hygiene requirements of infection control almost impossible in the informal settlements of rapidly ur-banising cities. To this, we must also add the impact of pandemic infection on those populations already experiencing a high poor-health burden and limited access to medical services. This creates a complex recipe for high infection and mortality rates. The pandemic has laid bare the fundamental inequalities between nations, one that has additionally been underlined by the inequality of access to vaccines.

Pandemic-related inequalities are not limited to low-income nations. In high-income countries, major differences in rates of infection and morbidity have been related to work characteristics, housing living density, and income levels. From the earliest days, the pandemic has hit poor communities hardest (Whitehead et al., 2021). In two of the nations that have been most severely impacted, the USA and the UK, early evidence pointed to up to 2.5 times the mortality rate for the poorest people, and official statistics have demonstrated a clear predominance of infection and mortality in the poorest neighbourhoods (Noppert, 2020), and among those in low status and low-income occupational roles (Windsor-Shellard and Rabiya, 2021). For front-line workers in the health, care, retail, and catering sectors, the inability to work at home and restricted work conditions, can create frequent social contact and opportunities for transmission. Historical health inequalities only add to the vulnerability of the poor (Patel et al., 2020).

The crisis we face is a truly global phenomenon of barely fathomable proportions. Its intermeshed and contested complexity is seldom addressed and even less understood. The pandemic has illustrated just how interrelated economies and societies are, and how each is also intimately connected to the environment. How a crisis is defined has major implications for the type of actions that are prescribed. For us, the place is the core lens through which the problems we face can be best understood. In regenerative development, the importance of place is widely understood, since all processes occurring are influenced by bioregional and cultural particularities (Foss, 2012). Thus, places are not introverted areas with boundaries, but 'fluid constructions of particular relational assemblages in certain environments and specific moments' (Massey, 1993, pp. 66–68). Place is the spatial location through which many of the solutions required to address the multidimensional crisis can be grounded, and the means by which sustainable and equitable transitions can be best democratised. It is not just a surface where ecological and inequalities play out; the way we see and treat place needs to change as a collective and potentially empowering place-making process.

The environmental and social crises

There has been a growing and dominant trend in policy discourse to discuss the environmental issues we face today as reductionist, independent, and discreet problems. Most commonly, governments refer to the 'climate crisis', which is largely defined as a CO_2 crisis, or, less frequently, as a 'nature crisis'. Our contention in this book is that, although there is certainly an urgent need to fundamentally redress the destructive relationship humans have with the atmosphere and with nature, to see these as somehow independent of each other or as the limit of harmful human impact on the world, seriously restricts the scale of the change we need to make to human civilisation. The 'environmental crisis' and the 'social crisis' are multifaceted and deeply interconnected. The scale of change needed cannot be understood through a series of silos that somehow reflect the policy and bureaucratic structures of modern governments or scientific disciplinary interests. Thus, politically symbolic climate change departments, green economy or green jobs strategies, or 'levelling up' programmes that seek to reduce the problems to administratively manageable proportions are doomed to fail. They fail because they do not grasp the far more fundamental nature of the change required, which cuts across governance and disciplinary boundaries.

There is a wide range of interconnected issues that comprise our global crisis, and it is worth briefly outlining these here. In doing so, we are keen to stress that the causes of each, and potential solutions, cannot be found in isolation:

- *Climate change* (the pollution of the atmosphere by greenhouse gasses). Since the Industrial Revolution, human impact has had the greatest impact on driving global climate change primarily by the burning of fossil fuels (coal, oil, and gas), which have increased so-called 'greenhouse gas' concentrations. These are now at their highest levels in 2 million years and emissions are continuing to rise. As a consequence, the UN Intergovernmental Panel on Climate Change AR6 Report (IPCC, 2021) warns that without deep reductions in GHG (Green House Gases) production, the Paris Agreement target of 1.5°C will not be met, with significant negative global impacts on weather, sea levels, biodiversity, and a wide range of other ecosystem impacts.
- *Stratospheric ozone depletion* (the depletion of the ozone layer by halocarbons pollution). This is perhaps one of the few examples where the action taken to phase out CFCs and other ozone-depleting substances following the Montreal Protocol on Substances That Deplete the Ozone Layer in 1987, which was ratified by all 197 UN member countries, is showing good results. The ozone layer is now expected to heal completely in the (non-polar) Northern Hemisphere by the 2030s, followed by the Southern Hemisphere in the 2050s and polar regions by 2060 (Nunez, 2019).
- *Air quality* (local and regional air pollution). Although human activity has caused air pollution as early as the Antiquity (through livestock domestication, metallurgy, and wood burning), it is the Industrial Revolution that severely altered air quality due to coal combustion, increasing emissions of SO_2, NO_2, NH_3, and smoke. By now although, most primary air pollutants seem to have reached a peak in Europe, North America, Japan, and China, yet governments should remain cautious and increase efforts to improve air quality. Although on a downward trend, high levels of PM2.5 persist and negatively affect the health of millions of people, and a lack of actions to reduce ammonia might further worsen feedback loops (Fowler et al., 2020). As with most sustainability issues, air quality has an unequal global distribution, with major discrepancies between countries, as well as within

them. Megacities from India, China, and some of the African countries continue being the most affected by poor air quality, and the WHO estimated that 4.2 million deaths occur as a result of ambient air pollution globally (WHO, n.d.).

- *Water quality* (coastal and sea water pollution). The primary causes of sea and coastal water pollution arise from chemicals and rubbish. Chemical pollution, from sources such as farm fertilisers, 'runoff' into water courses that ultimately end up in coastal and sea water. The impact of rubbish, most notably plastic, on coastal and sea water has been significant, and the UN Environment Programme (2019) estimated that 8 million tonnes of plastic waste enter the world's oceans every year, which is having a significant impact on marine life and biodiversity.
- *Fresh water scarcity* (scarcity in some areas and poor water resource management in others). UNICEF (2021) estimated that 1.42 billion people, including 450 million children, live in areas of high or extremely high water vulnerability. This has significant impacts on human health, and also on the natural environment. Climate change is one factor in this scarcity, but, so is poor water resource management, notably by agriculture, and much water is wasted through inefficiencies. Other causes such as land-use change, population growth, and urbanisation are placing significant demands on water resources and exacerbating the problem.
- *Land contamination* (land pollution). Land can be contaminated through a wide range of substances arising from human activities (including, although not limited to, industrial and agricultural), including: heavy metals (i.e., arsenic, cadmium, and lead)'; oils and tars; chemical substances and preparations (i.e., solvents); gases; and radioactive substances. Such contamination has adverse effects on human and environmental health. The European Environment Agency (2019) concluded that there is a lack of a comprehensive policy framework for protecting land in Europe, and this is likely to be the case elsewhere.
- *Deforestation* (from air to food, forests are critical to global ecosystems). After the oceans, forests are the second largest store of global carbon. They are home to over 50 percent of land-based species. Deforestation is occurring at an alarming rate, however, with two-thirds of global forest loss taking place in the tropics and sub-tropics. The WWF has estimated that over 43 million hectares were lost between 2004 and 2017 (an area it equates to roughly the size of Morocco) (WWF International, 2021).
- *Soil erosion and degradation*. The IPCC (2019) showed that land degradation is currently occurring over a quarter of the Earth's ice-free land area, and agriculture is one of the major drivers of degradation. Soil loss caused by conventional agricultural practices exceeds the rate of soil formation, leaving over 1 billion people worldwide vulnerable, as their livelihoods depend on natural systems with ever-reducing productivity. Nonetheless, sustainable land management practices, restoration, and rehabilitation technique could

improve the situation, offering co-benefits, which include climate change adaptation and mitigation (IPCC, 2019).

- *Land-use change and habitat loss.* Humans have been altering natural landscapes in various ways, from permanent changes induced by building settlements, to attempts to restore forests and ecosystems and reduce the negative ecological footprint. Recent studies estimate that 'about three-quarters of the Earth's land surface has been altered by humans within the last millennium' (Winkler et al., 2021, p. 2). More often than not, land-use changes have led to habitat destructions, meaning that certain areas lost entire ecosystems to make room for human needs such as agriculture, housing, or infrastructure. Current extinction rates have risen to 100–1,000 times the average/background rate (Dasgupta, 2021), determining biodiversity to decline faster than ever in human history.
- *Biodiversity loss* (relates to the genes, species, and ecosystems that are essential to all life on Earth). The UN Environment Programme Global Biodiversity Outlook 5 has concluded that 'the increasing impacts of land and sea use change, overexploitation, climate change, pollution and invasive alien species...are in turn being driven by currently unsustainable patterns of production and consumption, population growth and technological developments'. The impacts will, the Outlook concluded, '...have a particularly detrimental effect on indigenous peoples and local communities, and the world's poor and vulnerable...' (Secretariat of the Convention on Biological Diversity, 2020, p. 12).

The scale of the multifaceted environmental crisis is a huge challenge for humanity. It is compounded by an equally significant social challenge that is closely connected and has its origins in some of the same causes of our alienation from nature. We have also lost some of our human interconnectedness and sense of collective responsibility for the well-being of all members of our human community. The growth of welfare provision in the developed world following the Second World War and the development of supra-national organisations to address global social problems was founded on a broad collectivism that, whilst imperfect, did attempt to ameliorate the 'social problem' at a national and global level. Welfare provision based on collectivist and universalist principles has been replaced by punitive welfare regimes of restricted qualification and eligibility. The global collective responsibility has been eroded by reduced aid budgets and dissent within key global organisations, perhaps epitomised by the America First nationalism of the Trump administration and its withdrawal from critical international agreements and organisations.

The consequent scale of poverty and inequality presents us with a further challenge. These are not new issues for this century. For Therborn, 'poverty, inequality and social justice are ancient plagues of human kind' (Therborn, 2019, p. ix). There is a sense, however, of their deepening and widening as the welfare provisions of high-income countries and the capacity of supra-national organisations to resolve humanitarian crises have weakened, particularly over the last 20 years. This failure of collective responsibility for the disadvantaged can be seen in several trends evident in human well-being:

- *Rising poverty in high-income nations.* Although most evident in the UK, the general erosion of welfare rights in Europe has led to increased levels of relative poverty and associated patterns of social exclusion. This is evident for post-industrial, white working-class communities as well as immigrant and ethnic minority communities (Francis-Devine, 2021). For sections of the population social exclusion, particularly from the labour market, prevents participation as 'full citizens' in the political, cultural, and economic life of the societies they are members (Clasen, 2002).
- *Child poverty.* Child poverty in the UK has remained stubbornly high despite considerable policy focus, particularly during the Blair government's commitment to end child poverty (Piachaud and Sutherland, 2001). Currently, 4.3 million (31 percent) children in the UK live in poverty. For black and ethnic minority children, the proportion is 46 percent, whilst 49 percent of children in lone-parent families in the UK live in poverty.
- *Food poverty.* The problem of food poverty and food insecurity is illustrated by the growth in food banks and their increasing use by people experiencing poverty. UK food bank charity, the Trussell Trust, operated 65 foodbanks in 2010–2011, providing 65,000 food parcels. By 2019, this had grown to 1,300 foodbanks, providing 1.9 million food parcels. Research by the Trussell Trust and Herriot Watt University concludes that the primary cause of this exponential rise is 'limitations and changes in the social security system' (Bramley et al., 2021, p. 15).
- *Global poverty and Inequality.* For low-income nations, problems of poor health, hunger, and climate crisis are life-threatening for many of the world's population. In low-income nations, particularly in sub-Saharan Africa and Southern Asia, these are existential problems and are reflective of long histories of domination and exploitation by the high-income nations of the global north. Despite some evidence of 'convergence' (Hickel, 2017), 4 billion people live with an income of less than US$5 per day (Bangura, 2019).

This social crisis, along with the environmental crisis, is explored in this book to demonstrate their interconnectedness and to show that their respective resolutions can be found in shared measures and innovative methods of economic and social organisation, restoration, and regeneration. Throughout this book, we also investigate a range of concepts that underpin our thesis, but which are explored to greater degrees in one or more chapters. These include, but are not limited to, the following: economic growth and neoliberalism (Chapter 2); poverty (Chapter 3), inequality (Chapters 3 and 5); Anthropocene and natural rights (Chapter 4); and place-based theory and praxis (Chapter 7).

How might we address the multiple challenges?

As well as diagnosing the interconnectedness of the problems – the world as it is – this book will also advocate a set of proposed pathways to overcoming these. The central thesis of this book is that place offers a critical window on how to interrogate and conceive of possible resolutions to the multiple and overlapping crises. The place is the community we and our families live in, where we tend to socialise with our friends or exchange pleasantries with neighbours, where many of our day-to-day basic goods are obtained and services are met, and where most of us work. During the COVID-19 pandemic, in particular, many of us have re/discovered our immediate neighbourhoods, engaged with nature and wildlife on local walks and in our own gardens. Many of us have volunteered more, often in unstructured and informal ways, offering assistance to the more vulnerable in our communities (Legal and General, 2020). We may even have moved away from materialism and towards community focused and pro-environmentalism (Evers et al., 2021).

In the midst of growing uncertainties, *places can offer hope*. The place offers a way forward for local communities as the testbed for sustainability, and through this book, we offer our suggestions on how we might best realise this potential. Despite the fundamental challenges we face, we are not pessimists. We see in the world around us the stirrings of local action, popular movements, and theoretical models that seek positive change in the world. From the 'build back better' arguments emerging during the pandemic, to serious attempts to redefine economic relationships evidenced in models, such as 'doughnut' or well-being economics, we see the imaginings of possibilities that move beyond the 'there are no alternative' claims of the neoliberals. We suggest in Chapter 10 that these are examples of, and signal the need for, a return to more collective approaches. We have termed this 'regenerative collectivism', where the natural world is also a member of the collective. As we have experienced in the past, there are alternatives, and there will be new and as yet unimagined alternatives. To achieve them will be a social, political, and economic struggle, but it is not impossible. We hope that this book will add to the tools that are available to us, to collectively assure a future, not only for our species but also for the many others we currently threaten.

This book is divided into two parts. Part One provides a critical synthesis of ideas and key concepts essential for a sustainable and equitable transition. This book will argue for the need for a real and green 'new deal' and will identify what this should be like. It will suggest that a new economic order, whilst eventually inevitable, requires radical change that will not be easy but is essential given the current impasse, caused, not least by the conjunction of carbon-based, neoliberal capitalism in crisis and the multifactorial global ecological crisis. Chapter 1, therefore, considers the economy and what kind of new economic settlement is required. Chapter 3, meanwhile, reflects on place and social structures – the intersection of economy and society. It therefore examines the causes and effects of poverty, social exclusion, and rising levels of regional, national, and international inequalities. This chapter also highlights the lasting abhorrent impact of slavery, colonialism, and post-colonialism.

Chapter 4 looks at the environment and offers an understanding of the socio-natural relations that exist during the Anthropocene. It considers how social and natural rights have become more unequal and are being increasingly marginal-ised. It also offers a discussion on the long and continuing process of detachment of people and nature and argues that as people's rights to natures diminish, or are held by a smaller number of people, so economic and social marginalisation increases. This chapter therefore suggests that only a transformational environ-mental agenda can empower people to take back their natures. Such an agenda requires place-based action, together with help from broader governance to fa-cilitate change. Chapter 6 examines the role of culture in place. It uses food, which it argues is the essential cornerstone of culture, as a metaphor to explore sustainable development, environmental degradation, and health. It also outlines the role of culture, again through the lens of food, in reinforcing gender ine-quality, as well as economic and social inequality. Chapter 6 offers a conceptual synthesis of Part One and locates place as a central part of this. It then argues how the links to place practices and approaches may be achieved, serving as a bridge to the following chapters.

In Part Two, we present the results of our experiments in place through a series of 'Deep Place' case studies that help illustrate our approach, and which support the broader advocacy of this book is outlined in Part One. The Deep Place methodology outlined in Chapter 7 shows the approach we have taken in implementing this place-based research. The methodology itself is not intended to be overly prescriptive, although it does offer suggestions on how such research and community planning may be undertaken in different locations based on our experience. It seeks and formulates sustainable pathways that combine ways to tackle the place-based manifestations of the twin crises. Chapters 8 and 9 outline the findings of our Deep Place studies. These studies have so far been undertaken in four UK communities (Tredegar, Pontypool, Lansbury Park, and Llandovery) (Chapter 8), one in Australia (Muswellbrook), and one in Vanuatu (Freshwater) (Chapter 9). These case studies are presented to demonstrate how place-based action can act as a location for the kind of change necessary to ensure just and sustainable transitions may be advanced. Chapter 10 explores how we move towards a more 'regenerative' model of economic and social organisation that is protective of both people and planet, defined as a 'collective' with shared challenges and needs.

Before we begin this exploration, it is important to establish a nomenclature for the distinctions between countries that have benefitted from a history of colonialism and those for which it has frustrated and even reversed their histor-ical human, social and economic development. Fialho and Van Bergeijk (2017) recorded a proliferation of categories in the post-war period and note that cat-egorisation is a process dominated by supra-national organisations such as the World Bank and the United Nations. They conclude, 'ultimately, international organisations are extensions and, hence serve the interests and values of their most powerful member states' (p. 106). Categorisations that deploy normative terms, including 'developed' and 'undeveloped', can be seen to continue the hierarchical relationships of the period of direct colonialism and, in much the same way that will be identified in the discussion of poverty in the UK in Chapter 3, they imply failing on the part of the nation itself. The comparison between 'developed' and 'undeveloped' nations is difficult to escape. Such comparisons have informed multiple reports and international indices and are a shared terminology across governments, aid agencies, and NGOs. Although its use here is a consequence of that ubiquity, we must acknowledge the inability of the terminology to adequately address the heterogeneity between nations categorised similarly (Khukhar and Serajuddin, 2015). We also note the processes of 'undevelopment' in European societies, where rising levels of poverty and inequality have reintroduced hunger, challenges of poor housing, deteriorating health indicators, and falling life expectancy.

A note on the authors

The authors contributing to this book are committed socially and politically to the resolution of the sustainability crisis and the eradication of poverty and inequality. Indeed, this book is intentionally developed to contribute to the global discussions and debates that move us collectively in that direction. Consequently, we are 'positioned' in a number of ways in relation to the topics covered in this book. The 'subjectivity' contained in our perspectives and its influence on our views is very different from the assertion of a value-free, objective social scientist often presented by researchers. Indeed, as Letherby and colleagues have suggested, 'our descriptions of the world are always partial, selected and filtered by our perceptual apparatus, by the assumptions that we bring to our observations, and by the particular perspective or standpoint from which we view the world' (Letherby et al., 2013, p. 7). We believe that it is impossible to be free of values, or to suspend them for the time engaged in research activity. Our beliefs and values are a constituent part of our consciousness and our identity. They shape who we are and what we think. We ask readers to bear this in mind as they read and inform their own understanding of the ideas we present. As much as possible we present facts and statistics, but there are value judgements in selecting what to present, or even what the subject of our research should be.

Despite these important limitations to our intellectual neutrality, we are also academics and strive to be objective, to weigh arguments and counterarguments and discern where the weight of opinion lies among those who are experts in the field. We believe the perspectives outlined in this book avoid the extremes of the current debates. For example, on climate change, we are guided by the work and publications of the IPCC and the multidisciplinary community of scientists who contribute to our knowledge of the field. In our understanding of poverty, we are informed by data and opinion from key international agencies, including the

United Nations. We make no apology for not replicating and giving credibility to those who would deny the reality of climate change and who do not find issues with the current levels of poverty within and between nations.

Similarly, the various Deep Place studies contained in Part Two are subject to the positionality of the researchers who have conducted them and the institutional settings in which they have been resourced. These will have a degree of influence on the implementation of the method, as will the very different societies in which they have been conducted. The original Deep Place study of Tredegar was conducted by Adamson and Lang in 2014 whilst at the Centre for Regeneration Excellence Wales, a government-funded organisation entirely dependent on the continuity of those funds. The study sought to demonstrate the failure of many years of social policy design and delivery and the continued existence of the twin challenges of environmental crisis and deep poverty in the community identified. Two of the subsequent studies in the UK – Pontypool and Llandovery both undertaken by Lang in 2016 and 2019, respectively – were resourced by the Sustainable Place Institute at Cardiff University, a location with considerable academic reputation and the associated rigour of potential peer review and publication. The Lansbury Park study, undertaken by Adamson and Lang in 2017, was supported by the local authority Caerphilly County Borough Council, with the specific objective of developing a well-researched plan to work towards addressing the deep poverty experienced by the community.

The Australian and Vanuatu studies in contrast were resourced by Compass Housing, an Australian not-for-profit, Community Housing Provider with a strong social mission. The studies were conducted by Adamson in 2016 and 2019, respectively, whilst working as an employee in a senior role with the organisation and subject to the key performance indicators of an executive position. These studies were also focused on key organisational objectives, which limited their scope and scale to the identification of optimal delivery of specific programmes and projects within the organisation. We are particularly aware of the clear cultural differences between the white Caucasian first-world researchers and the participants in the Vanuatu study, the latter steeped in the traditions and worldview of a particular Pacific people (Kang, 2020).

We see these differences as an effective indication of the flexibility and adaptability of the Deep Place method. Whilst the context is important for the outcomes, the core methodology is common to all and has shown its capacity to inform different objectives. As we outline in greater detail in Chapter 7, we see the method as fluid and developmental, changing as new insights and methodical components are adopted by researchers to inform their specific locations, challenges, and objectives.

References

Bangura Y (2019) 'Convergence is not equality', *Development and Change*, 50(2):394–409. https://doi.org/10.1111/dech.12489

Bargain O and Aminjonov U (2020) 'Trust and compliance to public health policies in times of COVID-19', *Journal of Public Economics*, 192: 104316.

Bramley G, Treanor M, Sosenko F and Littlewood M (2021). *Sate of Hunger: Building the evidence on poverty, destitution, and food insecurity in the UK.* Heriot-Watt University and the Trussell Trust. Accessed May 2022: https://www.trusselltrust.org/wp-content/uploads/sites/2/2021/05/State-of-Hunger-2021-Report-Final.pdf

Clasen J (2002) 'Unemployment and unemployment policy in the UK: increasing employability and redefining citizenship', pp. 59–74 in: J Goul Anderson, J Clasen, W van Oorschot and O Halvorsen (eds.) *Europe's New State of Welfare. Unemployment, Employment Policies and Citizenship*. Bristol: Policy Press.

Dasgupta P (2021) *The Economics of Biodiversity: The Dasgupta Review – Abridged Version*. London: HM Treasury.

European Environment Agency (2019) *The European Environment – State and Outlook 2020: Knowledge for Transition to a Sustainable Europe*. Luxembourg: European Union. Accessed 24 January 2022, from https://www.eea.europa.eu/signals/signals-2020/articles/land-and-soil-pollution

Evers N F G, Greenfield P M and Evers G W (2021) 'COVID-19 shifts mortality salience, activities, and values in the United States: big data analysis of online adaptation', *Human Behavior and Emerging Technologies*, 3: 107–126.

Fialho D and Van Bergeijk P A G (2017) 'The proliferation of developing country classifications', *Journal of Development Studies*, 53(1): 99–115. https://doi.org/10.1080/00220388.2016.1178383

Foss (2012) *What Is Regenerative Development?* Accessed 7June 2016: http://www.urban-thriving.com/news/what-is-regenerative-development/

Fowler D, Brimblecombe P, Burrows J, Heal M R, Grennfelt P, Stevenson D S, Jowett A, Nemitz E, Coyle M, Lui X, Chang Y, Fuller G W, Sutton M A, Klimont Z, Unsworth M H and Vieno M (2020). 'A chronology of global air quality', *Philosophical Transactions. Series A, Mathematical, Physical, and Engineering Sciences*, 378(2183): 20190314. https://doi.org/10.1098/rsta.2019.0314

Francis-Devine B (2021) *Poverty in the UK: Statistics. Commons Library Research Briefing No 7096 (Issue October)*. London: House of Commons library. https://doi.org/10.4324/9781843140184-8

Hickel J (2017) 'Is global inequality getting better or worse? A critique of the World Bank's convergence narrative', *Third World Quarterly*, 38(10): 2208–2222. https://doi.org/10.1080/01436597.2017.1333414

IPCC (2018) *Summary for policymakers. In: Global Warming of 1.5°C*. Accessed 24 January 2022: https://www.ipcc.ch/site/assets/uploads/sites/2/2019/05/SR15_SPM_version_report_LR.pdf

IPCC (2019) *Climate change and land. An IPCC special report on climate change, desertification, land degradation, sustainable land management, food security, and greenhouse gas fluxes in terrestrial ecosystems*. Accessed 22 December 2021: https://www.ipcc.ch/srccl/

IPCC (2021) 'Summary for Policymakers', in: V Masson-Delmotte, P Zhai, A Pirani, S L Connors, C P an, S Berger, N Caud, Y Chen, L Goldfarb, M I Gomis, M Huang, K Leitzell, E Lonnoy, J B R Matthews, T K Maycock, T Waterfield, O Yelek i, R Yu and B Zhou (eds.) *Climate Change 2021: The Physical Science Basis. Contribution of Working Group I to the Sixth Assessment Report of the Intergovernmental Panel on Climate Change*. Cambridge: Cambridge University Press. https://www.ipcc.ch/report/ar6/wg1/downloads/report/IPCC_AR6_WGI_Citation.pdf

Kang J (2020) 'Confronting shifting identities: reflections on subjectivity in transnational research', *The Qualitative Report*, 25(4): 937–946.

Khukhar T and Serajuddin U (2015) *Is the term 'developing world' outdated*. https://www.weforum.org/agenda/2015/11/is-the-term-developing-world-outdated/

Legal and General (2020) *10 million Brits volunteering as the nation unites in the Isolation Economy*. Accessed 21 January 2022: https://group.legalandgeneral.com/en/newsroom/press-releases/10-million-brits-volunteering-as-the-nation-unites-in-the-isolation-economy-says-legal-general

Letherby G, Scott J and Williams M (2013) *Objectivity and Subjectivity in Social Research*. London: Sage.

Marmot M and Bell R (2012) 'Fair society, healthy lives (Full report)', *Public Health*, 126(SUPPL.1): S4–S10. http://dx.doi.org/10.1016/j.puhe.2012.05.014

Massey (1993) 'Power geometries and a progressive sense of place', pp. 59–69 in: J Bird, B Curtis, T Putnam and L Tickner (eds.) *Mapping the Futures: Local Cultures, Global Changes*. London: Routledge.

Maxmen A (2021) 'US COVID origins report: researchers pleased with scientific approach', in: *Nature* (27 August 2021). https://doi.org/10.1038/d41586-021-02366-0

Noppert G (2020) 'COVID-19 is hitting black and poor communities the hardest, underscoring fault lines in access and care for those on margins', *The Conversation*, 9 April 2020.

Nunez C (2019) 'Ozone depletion explained', *National Geographic*. Accessed 21 January 2022: https://www.nationalgeographic.com/environment/article/ozone-depletion]

Patel M, Lee S I, Levell N, Smart P, Kai J, Thomas K and Leighton P (2020) 'An interview study to determine the experiences of cellulitis diagnosis amongst health care professionals', *BMJ Open*, 10: 034692. https://doi.org/10.1136/ bmjopen-2019-034692

Piachaud D and Sutherland H (2001) 'Child poverty in Britain and the new labour government', *Journal of Social Policy*, 30(1): 95–118. https://doi.org/10.1017/s004727940000619x

Rulli M, D'Odorico P, Galli N and Hayman D (2021) 'Land-use change and the livestock revolution increase the risk of zoonotic coronavirus transmission from rhinolophid bats', *Nat Food 2*. https://doi.org/10.1038/s43016-021-00285-x

Rustin M and Chamberlayne P (2002) 'Biography and social exclusion in Europe: experiences and life journeys', pp. 1–21 in: P Chamberlayne, M Rustin and T Wengraf (eds.) *Biography and Social Exclusion in Europe: Experiences and Life Journeys*. Bristol: Policy Press.

Secretariat of the Convention on Biological Diversity (2020) *Global Biodiversity Outlook 5 – Summary for Policy Makers*. Montréal.

Therborn G (2019) 'Preface: the terrifying convergence of the three worlds of the social question', pp. ix–xii in: J Breman, K Harris, C Lee and M van der Linden (eds.) *The Social Question in the Twenty-First Century: A Global View*. Oakland: California University Press.

UNICEF (2021) *Reimagining Wash: Water Security for All*. New York: UNICEF. Accessed 24 January 2022: https://www.unicef.org/media/95241/file/water-security-for-all.pdf

United Nations Environment Programme (2019) *Addressing Marine Plastics: A Systemic Approach – Recommendations for Action*. Nairobi: UNEP. Accessed 24 January 2022: https://www.unep.org/resources/report/addressing-marine-plastics-systemic-approach-recommendations-actions

Whitehead M, Taylor-Robinson D and Barr B (2021) 'Poverty, health, and covid-19', *BMJ*, 372: 376. http://dx.doi.org/10.1136/bmj.n376

WHO (n.d.) *Air pollution*. Accessed 25 January 2022: https://www.who.int/health-topics/air-pollution#tab=tab_1

Windsor-Shellard B and Nasir R (2021) *Coronavirus (COVID-19) related deaths by occupation, England and Wales*. ONS. Accessed 20 January 2022: https://www.ons.gov.

uk/peoplepopulationandcommunity/healthandsocialcare/causesofdeath/datasets/cor
onaviruscovid19relateddeathsbyoccupationenglandandwales

Winkler K, Fuchs R, Rounsevell M and Herold M (2021). 'Global land use changes are four times greater than previously estimated', *Nature Communications*, 12. https://doi. org/10.1038/s41467-021-22702-2

WWF International (2021) *Deforestation Fronts: Drivers and Responses in a Changing World*. Gland: WWF. Accessed 24 January 2022: https://wwfint.awsassets.panda.org/downloads/ deforestation_fronts___drivers_and_responses_in_a_changing_world___full_ report_1.pdf

2

GREEN DEALS AND A NEW ECONOMIC SETTLEMENT

This chapter considers two interrelated problems. First, the contradictions of current policy responses to the environmental crisis within the realm of economic policymaking. Second, the inherent 'lock-in' problems of the current global economy – if consumption declines in one place there are implications for those places producing the goods that are currently consumed. These problems lay bare the difficulties of designing and managing a transition to a more sustainable economic global model. In considering the fundamental weaknesses and destructive elements of the current model – such as the centrality of growth and the thesis of agglomeration, as well as the commodification, financialisation, and externalisation of environmental costs – we do not underplay the difficulties of a successful transition. This chapter firstly addresses the hegemony of neoliberalism, its intimate relationship with growth, and its enduring hold on political and economic elites. It then outlines the growing arguments in favour of change and the options that might support the economic transition, before considering how positive change should be measured and directed. We conclude that the type of economies required for sustainable place-making and global good, need to be distributed, non-extractive, socially just, and cooperative, rather than competitive.

Growth and neoliberalism

Growth is central to neoliberalism. Whereas many other ideologies pursue growth as a means to an end, neoliberalism appears to pursue growth as an end in itself – it is its *raison d'être*. It is, for example, possible to conceive of socialism without growth, but neoliberalism without growth seems like an impossibility. Since the 1970s when a general faith in Keynesianism declined, most notably in

DOI: 10.4324/9781003221555-3

the USA and the UK, neoliberalism has been hegemonic in economic policy discourse. Yet, we seldom pause to reflect upon the nature of neoliberalism. So pervasive has it become, we even fail to notice that it's actually there – not because its effects are benign, but because we have been collectively conditioned into assuming a certain inevitability. The term TINA or 'there is no alternative' was first used by Herbert Spencer in the 19th century, but, via Margaret Thatcher in the UK, it became a core slogan of neoliberalism. When coupled with the *end of ideology* assertions from Daniel Bell (1960) and *end of history* claims by Francis Fukuyama (1992), the triumph of the neoliberal model was seen as absolute. Neoliberalism subsequently became the only show in the economic establishment's town and, given their intimate relationship, growth has become the central purpose of economic policy.

Neoliberalism, of course, emerged long before the 1970s. Hayek's 1944 work *The Road to Serfdom* was deeply influential on future generations of neoliberals such as Milton Friedman, who was himself most closely associated with a particular brand of neoliberalism, monetarism, which dominated from the late 1970s. Monetarism is concerned with the supply of money in circulation and argues that an excess of money supply leads to inflation. From this perspective, the amount of money in circulation should only be increased by small, incremental amounts and closely tied to increases in the overall size of an economy. Neoliberalism, at its core, believes that competition is the defining characteristic of human relationships. It redefines citizens as individual consumers, where the free market is far more efficient and advantageous than any state planning could ever be. By extension therefore, taxation should be minimised, public services should be privatised, and it is believed that welfare and employment rights and labour self-organisation endanger the benefits of the 'free' market. This economic justification is reinforced by a political commitment to the small state to avoid the 'slippery slope' to totalitarianism identified by Hayek. This configuration of political and economic beliefs has created a hegemonic structure that socialism and other economic alternatives have struggled to counter with any argument that matches the simplicity and coherence of the neoliberal discourse.

Since mid-century growth ran out of steam in the 1970s, governments increasingly followed the neoliberal logic, reducing production costs by limiting wages, cutting social benefits and wider public services, weakening trade unions, and reducing employment standards. Punitive welfare regimes have further disciplined the workforce to accept increased casualisation and deregulation of labour and further depression of wage levels (in Chapter 3, we discuss the implications of these measures for the rise of poverty in the UK). At the same time, consumption was stimulated via a supply of consumer credit and more readily available mortgages (Kallis et al., 2020, pp. 28–29). In the UK, the Thatcherite council house 'right to buy' was sold as an extension of democracy, but was very much part of the neoliberal agenda to strip the state of public service responsibilities and – in this case local authorities – of power. The UK was, as former Conservative minister Ian Gilmour (1992) pointed out, *dancing with dogma*.

For us, neoliberalism is an economic model that rewards capital and subjugates labour. Consumers are sold the fantasy that a product or service will improve their lives. Increased consumption is funded by personal debt and, consequently, not only are more goods and services sold, but the loaned money earns interest on debt that is incurred. As the poorest in society pay the highest rates of interest through products such as 'pay day loans', they increasingly struggle to make ends meet. There is a direct consequence of these seismic shifts away from the post-war consensus and its distributive welfare state. One barometer of the growth of poverty is the substantial rise in a number of charitable food banks in the USA, Australia, Canada, the UK, and elsewhere, to support families that are no longer able to feed themselves as they have become caught in a vicious cycle of debt and precarious low paid employment. As of February 2021 in the UK, for example, there were over 2,200 food banks (Tyler, 2021). Meanwhile, capital earns interest on the loans made to purchase the disposable products or services it sells through increasingly pervasive advertising – for capital it's a win-win!

Employment is not necessarily a route out of poverty. The Joseph Rowntree Foundation identified a transition in the first decade of this century in Wales, for example, from economic *in*activity as the primary cause of poverty to the ascendency of in-work poverty (JRF, 2013). Stagnant wage growth has been a global feature of many economies since the 2008 financial crisis (Mason, 2015) and is a significant characteristic of poverty. In the UK, as elsewhere, the rise in the rate of in-work poverty has occurred despite a rising employment rate (JRF, 2021). As will be discussed in Chapter 3, although part of the reason lies outside of the labour market – for example, benefit cuts and rising living costs – much of this is driven by increasingly precarious patterns of employment. Increasing the employment rate is, therefore, insufficient without a critical examination of the type and quality of jobs that are created and their remuneration levels.

Tim Jackson (2021) suggested that

> for workers, wages are their principal source of income…. since labour is a cost to production and capitalists are motivated by profit, they will tend to do whatever they can to increase labour productivity: that is, to reduce the cost of labour.
>
> *(p. 30)*

Under the direction of neoliberalism, employment has become increasingly precarious, poorly paid, and unrewarding. Nevertheless, the central planks of regional economic strategies tend to play to the notion that the so-called 'inward investment' of capital should be encouraged and supported, through renewed infrastructure, the development of investment-ready skills bases, and often direct financial inducements. National, regional, and local taxpayers are subsidising capital to produce goods or deliver services to sell to us, based on an indoctrinated belief that all we need are jobs and growth. In the race to attract investment, countries and regions are pitted against each other.

Consistent with the neoliberal philosophy that competition is the defining characteristic of human interaction, places are driven to compete. This can be seen in the UK's transition from Regional Development Agencies, focused on regional spatial planning, to the multiple City Deals promoted by the recent conservative governments, which have provoked even neighbouring cities into competition against each other. Cities with larger populations of 'creative classes' are, it is espoused by a particular strand of neoliberalism, likely to be the most competitive in the race to secure investment (Engelen et al., 2016). Richard Florida (2005), who is highly influential in this approach, emphasised the need for cities to attract the creative classes by building infrastructure and the cultural trappings of the affluent society. Glaeser (2011), another protagonist for the cause, argued that there should be no support for struggling places and declining industries as this distorts the free market.

The so-called New Economic Geography (Hildreth and Bailey, 2013) has also gained significant traction in economic policy discussions. Informed by economists like Paul Krugman, this promotes the merits of spatially agglomerating industries (Krugman, 1998) as a means to increase knowledge spill-overs to drive economic growth (Pike and Tomaney, 2009). Redistributive economic policy, from this perspective, undermines agglomeration (Martin, 2008), and therefore so-called 'resource depleted' places should be allowed to decline, irrespective of the human costs. The agglomeration logic has also been supported by others who believe the UK should stop jam-spreading funds (Overman, 2013) by investing a little bit everywhere to rebalance territorial inequalities. Instead, the governments should look to improve conditions in more successful cities, allowing any additional growth to eventually trickle down or out.

Paradoxically, many of those who support this agglomerative and place competitive agenda also continue to support often significant public sector infrastructural investment (or direct financial inducements) to create the conditions for competitive success. The protagonists calling for reduced public expenditure in 'failing places' tend not to ask if the much applauded 'successes' of places like London (little is said about the extremely high poverty rates in such places) can, at least in part, be attributed to higher public sector expenditure over a prolonged period. This has been described as a process of 'cumulative causation' (see: O'Hara, 2008). Neoliberal economic policy is, incongruously, heavily subsidised by the state.

Meanwhile, the economy and Gross Domestic Product (GDP) have become virtually indistinguishable. As David Pilling (2018) pointed out, 'before the invention of GDP, it was pretty difficult to define what an economy was, even if you wanted to' (p. 10). Growth (as measured by increases in the size of GDP – defined as the value of goods and services produced during a given period) and economic success are inseparable and, consequently, growth is *the* standard against which governments are judged. This perspective has become entrenched because of an underlying belief that growth produced some worthwhile social outcomes during the mid-20th century (Spretnak and Capra, 1986; Bleys and Whitby, 2015), and it therefore continues to be assumed that everyone benefits from a growing economy (hence the current UK Government policy discourse of 'levelling up').

Correlation does not, of course, imply causation. Thomas Piketty (2014) has conclusively debunked the myth that the mid-20th century reduction in income gaps was the result of economic growth. Instead, he proved that this was a result of the destruction of wealth during the Great Depression and World War Two, and significant post-war egalitarian policies adopted by successive governments in Europe and North America. Moreover, since 1980 when neoliberalism and monetarism have been in the ascendance, growth has actually been accompanied by greater inequality, not less. Neoliberal economics, it seems, has created a particular form of capitalism that has worked exceptionally well for some, but continues to fail the majority of people (Jackson, 2021). The 'trickle-down effect' is as mythical as the tooth fairy, both are sold to us in an effort to take the pain away, but both are socially constructed fantasies.

Changing direction

It has been argued that the failure of growth to deliver positive outcomes to a broad spectrum of society, which we discuss in Chapter 3, and its destructive environmental and cultural impacts, which we discuss in Chapters 4 and 5, may be explained by currently adopted measurements. For Fioramonti (2017), 'common sense [suggests]…growth happens when we generate value that wasn't there before'. This may, for example, include education, improvements in public health, or food preparation. If these activities generate costs, Fioramonti suggests, these should be deducted from the value created, and consequently growth therefore 'equals all gains minus all costs.' The currently applied model of economic growth, however, does the opposite of this, as nowhere in the existing growth equation do detrimental costs feature. As currently measured, growth is irrelevant to discussions about well-being.

Kevin Gulliver (2017) developed the idea of the 'human city'. This is not something restricted to cities, as the name might suggest, but rather is defined as:

> a geographically defined settlement – a city, town or village – where policies, practices and initiatives are enacted to ensure the best of human endeavours can flourish and where citizens and communities can shape an equitable, affordable and a shared society.
>
> *(p. 13)*

In the human city, localised economies with a high incidence of small enterprise and local and/or community ownership are more cost-effective and support better public health and well-being. Effective local control of grounded assets is a requisite for such a settlement, which has been emphasised by Love (2017) in their pursuit of asset-based community development in Glasgow, Scotland, where they have sought to identify and more effectively harness community skills and strengths. Although we agree that such an approach can make communities stronger and deliver a greater well-being dividend, the approach is not captured in the current growth-based model. A radical change of perspective is therefore required to enable such approaches to realise their true potential.

The UN Stiglitz report (2009) published in the immediate aftermath of the global financial crisis, suggested that 'reform of the international system must have, as its goal, the improved functioning of the world's economic system in support of the global good'. This, the report argued,

> ...entails simultaneously pursuing long-term objectives, such as sustainable and equitable growth, creation of employment in accordance with the "decent work" concept, responsible use of natural resources, reduction of greenhouse gas emissions, and more immediate concerns, including addressing the challenges posed by the food and financial crises and global poverty.
>
> *(p. 13)*

If such sustainability ambitions are to be realised, however, fundamentally contradictory issues concerning the impact of growth must be addressed.

The Ellen MacArthur Foundation, which was established to promote the development of the 'Circular Economy', argues that we should 'decouple' economic value creation (growth) from resource consumption. It argues that reducing the resources and fossil energy consumed per unit of economic output will only delay the inevitable, and therefore calls for radical changes and the development of a 'circular economy' (Ellen MacArthur Foundation, 2016). There has been significant interest from governments, NGOs, and companies in these ideas. Despite the clearly well-intentioned motivations behind the proposals of the Ellen MacArthur Foundation, however, concerns have been raised about the ability to decouple economic value creation from resource consumption.

Zink and Geyer (2017) suggested that the economics of the circular economy might partially offset the potential benefits, as ultimately lower per-unit production impacts also cause increased levels of production. They describe this effect as 'circular economy rebound'. A similar point was made by Tim Jackson (2009), who argued that

> those who promote decoupling as an escape route from the dilemma of growth need to take a closer look at the historical evidence...improvements in energy (and carbon) intensity...were offset by increases in the scale of economic activity....
>
> *(p. 86)*

Nevertheless, the prospect of 'decoupling' growth from the negative environmental impact has gained significant traction with governments and policymakers.

The decoupling agenda was adopted as a policy objective by the OECD in 2009, leading to its green growth strategy (OECD, 2011). Similarly, the European Commission has adopted decoupling by stating 'through greater efficiency and better use of natural resources, we can break the old link between economic growth and environmental damage' (European Commission, 2001, p. 3). Later, the United Nations Environment Programme's strategy on the green economy gave similar commitments (United Nations EP, 2011). Even the World Bank has claimed its support for 'green growth' with a commitment to decouple (World Bank, 2012). Uniting these policy commitments is a belief in the possibility of growth, without further environmental degradation. Decoupling appears to offer a 'get out of jail free' card, but there are concerns about the logic that underpins this 'green growth' model.

Parrique and colleagues (2019) suggested that green growth wrongly assumes that sufficient decoupling can be achieved through increased efficiency without the need to limit growth. This logic, they suggest, '...relies on the assumption of an absolute, permanent, global, large and fast enough decoupling of economic growth from all critical environmental pressures' (p. 4). They intimate that decoupling needs to be sufficiently large in affluent countries to free the global ecological space necessary for production and consumption in places where basic needs currently go unmet. They argue that not only is there no empirical evidence to identify such a decoupling taking place, but such decoupling is unlikely to happen in the future for seven reasons:

1 *Rising energy expenditures* – 'when extracting a resource, cheaper options are generally used first, the extraction of remaining stocks then becoming a more resource and energy-intensive process'.
2 *Rebound effects* – 'efficiency improvements are often partly or totally compensated by a reallocation of saved resources and money to either more of the same consumption, or other impactful consumptions. It can also generate structural changes in the economy that induce higher consumption'.
3 *Problem shifting* – 'technological solutions to one environmental problem can create new ones and/or exacerbate others'.
4 *The underestimated impact of services* – the service economy can only exist on top of the material economy, not instead of it'.
5 *Limited potential of recycling* – 'recycling rates are currently low and only slowly increasing, and recycling processes require a significant amount of energy and virgin raw materials'.
6 *Insufficient and inappropriate technological change* – technological progress is not targeting the factors of production that matter for ecological sustainability and not leading to the type of innovations that reduce environmental pressures'.
7 *Cost shifting* – 'what has been observed and termed as decoupling in some local cases was generally only apparent decoupling resulting mostly from an externalisation of environmental impact...'. (pp. 4–5)

In our previous work (Lang and Marsden, 2021), we suggested that a further major trend could be added to this list: the *continued privatization of the 'commons'*. We suggested, that through advances in extraction, data mining and bio-physical processes – what Christophers (2018) calls the "new enclosures" – this relies upon a neoliberal state willing to grant more privatized enclosures as indeed corporate capitalist development confronts limits on extraction not least from environmental resistance movements.

This reduces nature-based common-pool resources. In the UK over the past 30 years, the amount of publicly owned urban and rural land has been reduced by 50 percent. In urban areas, pseudo-public spaces – spaces accessible to the public, which look like public land, but which are privatised – have been mushrooming, redefining acceptable behaviour based on the landowner's rules (Schenker, 2017). We discuss these issues further in Chapter 4.

These are important warnings, as ecological modernisation based upon green growth will not deliver as long as it is a compromise between two dominant discourses: neoliberal and multi-level states that finance the agglomerative growth with all its defective 'trickle down' assumptions; and a reliance upon restrictive definitions of growth and competitiveness that exclude more embracing concepts of asset wealth and well-being (Lang and Marsden, 2021). This is further complicated by the surrendering of democratic control by national governments to transnational capital through unilateral and multi-lateral trade deals, which can sanction them if returns to capital are diminished (for discussion see: Freeman, 2020). To pursue sustainable well-being, states must abandon the faith that a dual model of spatial and eco-economic development can be built, which supports and leaves the dominant growth model largely intact. This brings us to our concerns relating to Green New Deal.

Green New Deals

There are, what Mastini and colleagues (2021) described as, 'two master narratives on climate change': degrowth and Green New Deal. Advocates of Green New Deal maintain that investments in renewable energy will inevitably grow related activities, have spill-over effects and consequently stimulate the economy. The resultant economic growth will then increase the revenues available for clean energy investment. This places Green New Deal at odds with degrowth. There have, in effect, been two distinct periods where Green New Deal has appeared in public discourse: the first between 2007 and 2010, and the second, more recently, from about 2019 onwards. In each period, Green New Deal has had a slightly different flavour.

During the first phase, Green New Deal in the USA was very much tied to a broad range of programmes and projects designed to revitalise the American economy. Barrack Obama's America Recovery and Reinvestment Act should be seen in this context. Within a total stimulus amounting to 976 billion US dollars, however, just 117 billion US dollars were oriented specifically towards energy efficiency and renewable energy (Mastini et al., 2021). Although clearly a sizable sum was therefore allotted to green measures, the emphasis was still very much on the broader stimulus of the American economy. In the immediate aftermath of the financial crisis, the initial response of various governments and international institutions was similar to that of the USA: a proportion of green investments as part of a much more substantial fiscal stimulus package.

In the UK, New Economics Foundation (2008) published the first Green New Deal of the UK Green New Deal Group, which called for an initiative for economic and environmental transformation, and the re-regulation of finance and taxation and major government investment in renewable resources. The United Nations Environment Programme policy brief on Green New Deal in March 2009, recommended that governments allot expenditure equivalent to 1 percent of GDP on green initiatives, although in practice the G20 group overall spent just 0.8 percent of GDP on green measures (Mastini et al, 2021). In 2010, neoliberalism reasserted itself and the global economic consensus shifted from stimulus to austerity and Green New Deal quickly disappeared from political discourse.

In March 2019, *House Resolution 109* was passed in the US House of Representatives. The resolution stated that a Green New Deal should be instigated in order to address a climate crisis and an economic crisis of wage stagnation and growing inequality. This, arguably, began the second phase in Green New Deal, one which appears to have more echoes of Franklin D Roosevelt's New Deal, after which it is named, than the initial phase, as it now seeks to address both a social, as well as an ecological crisis. The central features of this phase two Green New Deal include: public ownership of energy utilities, social and labour reforms, and a job guarantee. Whilst House Resolution 109 does not specifically mention growth as a policy objective, it would nevertheless seem implicit given that the text refers to 'spur economic development' and 'grow domestic manufacturing' as objectives. Green New Deal advocates suggest that growth can be beneficial for environmental action, as they proposed the additional tax taken on a growing economy could be used to fund green initiatives, whereas degrowth advocates suggest that growth makes overcoming environmental degradation more difficult, if not impossible.

Budgeting for well-being

One of the arguments put forward by critics of degrowth is that an economy that is shrinking results in a smaller tax take from which to fund green measures. In addition, Robert Pollin (2019), for example, suggested '…any global GDP contraction would result in huge job losses and declines in living standards for working people and the poor'. Moreover, he says, '…the prospects for reversing inequalities in all countries would be far greater when the overall economy is growing, than when the rich are fighting everyone else for shares of a shrinking pie' (pp. 317–318). This would seem, however, to imply a continued faith in the existence of a trickle-down effect and a continued faith in everyone benefiting from a growing economy. It also negates the possibility of making additional changes to counteract these forces and appears to run counter to Thomas Piketty's evidence, that the destruction of wealth during the two world wars and strong egalitarian policies following WW2 were actually the cause of reduced inequalities in the period up to the 1970s rather than growth. In fact, evidence shows that in countries like the USA, economic growth and life expectancy are now disconnected – while the first continues to rise, the latter has been decreasing (Solly, 2018). In the UK too, research based on ONS data showed that life expectancy across the UK, for both women and men, started decreasing from 2015 onwards, arguing that this was a result of public spending cuts (Dorling, 2019).

For us, tackling the social or environmental crises does not require economic growth. This is just as well, as there is evidence that the rate of growth in the global economy is generally slowing down (Gopinath, 2019). Mastini and colleagues (2021) suggested three possible strategies for funding public investments without relying on economic growth:

- Public expenditures could be reallocated away from socially and environmentally harmful sectors or gleaned from the expected positive effects of the ecological transition.
- Governments could tap into private and corporate savings by means of progressive taxation, which would have an environmental use and a social inequality impact.
- Money creation could be decommodified and reorganised as a common good. A sovereign money system would entail debt-free money creation on the part of a country's central bank.

This last point would appear to be consistent with the approach advocated in Modern Monetary Theory (see: Whittaker, 2020).

Aparna Mathur (2019) argued that a carbon tax could be a central legislative reform to address both social inequality and environmental crisis. She suggested that carbon pricing is an effective mechanism to achieve emissions reduction targets, but whereas a carbon tax might help achieve Green New Deal goals on emissions targets, its regressivity could potentially exacerbate inequality issues as poorer households are likely to pay a larger proportion of their annual income on a carbon tax than the wealthy. Nevertheless, she said, if such a carbon tax were accompanied by other transfer programme changes, then this might be overcome.

How do we measure progress?

The preceding discussion leads us to a rather fundamental question that is seldom asked in mainstream economic policy, even though it is increasingly being asked by others such as the international group Economy for the Common Good: *what should our economy be for and how should we design and measure progress toward it?* If we accept that GDP growth is not an indicator of well-being, and the current economic order delivers for neither social nor ecological good, then an alternative system and measurement become necessary. One attempt to rethink the economic system and measure progress towards a new system is Kate Raworth's (2017) 'Doughnut Economics'. Raworth's doughnut shape diagram, used to illustrate her thinking, includes two concentric rings. The lower ring is the social foundation, which consists of critical human deprivations such as hunger and illiteracy. The upper ring represents the ecological ceiling and consists of critical planetary degradation such as climate change and biodiversity loss. Between the two rings is Raworth's doughnut, the place she says in which we can meet the needs of all within the means of the planet. In her model, there are various social and ecological indicators to measure progress.

There are seven themes around which Raworth designed her 'doughnut economics', these include:

- Change the economic goal.
- See the big picture – invite new narratives about the power of the market, the partnership of the state, the core role of the household, and the creativity of the commons.
- Nurture human nature – we are social, interdependent, approximating, fluid in values, and dependent upon the living world.
- Get savvy with systems – it's time to stop searching for the economy's elusive control levers and start stewarding it with an ever-evolving complex system.
- Design to distribute – inequality is not an economic necessity; it is a design failure.
- Create to regenerate – we need a circular not linear economy (a point further discussed in Chapter 5).
- Be agnostic about growth – we need economies to make us thrive, whether or not they grow.

Raworth's analysis is part of a growing body of work that is fundamentally rethinking our economic systems and experimenting with alternative measures of success. In the USA, some of the individual states have been using the Genuine Progress Indicator (GPI) as an alternative to GDP to measure whether economic progress results in sustainable prosperity. GPI accounts for effects of income inequality on well-being; values volunteering and education; and subtracts the negative effects of crime, unemployment, and pollution (Kubiszewski et al., 2013). Elsewhere, Bhutan has used a Gross National Happiness indicator since 1972. Although these measures do not formally abandon growth or adopt a degrowth agenda, they do show an increasing recognition that GDP growth has had its day (see Lang and Marsden, 2018).

The Happy Planet Index was constructed by the New Economics Foundation to identify global well-being by country. The index includes measures of well-being, life expectancy, inequality, and ecological footprint to give an overall rank. The results show that a country's GDP is no guarantee of national happiness. It is not perfect, but it is an attempt to move away from using growth as an indicator of well-being. In 2019, New Zealand formally abandoned GDP as a primary measure of economic success to concentrate on issues it felt were more important to well-being, and this perhaps explains why the country was able to take such decisive action to stop the spread of COVID-19 – it saw no economic difficulty in 'locking down' even before it had seen a single death (see discussion in Attenborough, 2021). Although there are indications that in the post-COVID-19 period New Zealand's previously stated ambitions may be more balanced and less ambitious (McClure, 2021).

In Wales, the devolved legislature passed national well-being legislation in 2015. The *Well-being of Future Generations (Wales) Act* places a legal duty on devolved public bodies to safeguard the economic, social, environmental, and cultural well-being of Wales. The Act contains seven Well-being Goals that public bodies must take all reasonable steps to support, including: Prosperous Wales, Resilient Wales, Healthier Wales, More Equal Wales, Wales of Cohesive Communities, Wales of Vibrant Culture and Thriving Welsh Language, and A Globally Responsible Wales. The Act also contains a Sustainable Development Principle and public bodies need to show that their actions and policies are consistent with 'five ways of working': Long-term, Prevention, Integration, Collaboration, and Involvement.

The Welsh Future Generations Act in many respects embodies and extends many of the wider Sustainable Development Goals (SDGs) developed by the United Nations. In particular, SDG 8 (on creating decent working conditions and sustainable economic growth) and SDG 12 (concerning responsible production and consumption). The Welsh Government's recent economic strategy (Welsh Government, 2017), however, has also outlined an 'Economic Contract', which sought to 'stimulate growth, increase productivity and make Wales fairer and more competitive' (p. 8). Moreover, it requires businesses asking for Welsh Government support to demonstrate:

- Growth potential (contribution to employment, productivity, or multiplier effects through the supply chains).
- Fair Work (as defined by a Fair Work Board).
- Promotion of health (emphasis on mental health, skills, and learning in the workplace).
- Progress in reducing carbon footprint.

On the reducing carbon footprint condition, the strategy says:

> we need to transform our economy to use fewer resources per unit output and keep products and material resources in high value use for as long as possible. We are helping businesses through the transition to a low carbon economy and are coupling this with a move to a circular economy.
>
> *(p. 25)*

The Welsh economic strategy demonstrates a faith in the ability of the circular economy to ensure economic growth can continue to be pursued, without further environmental degradation. The Welsh Government would therefore appear to have adopted faith in the decoupling thesis. For us, although we recognise that well-being legislation in Wales is truly pioneering, it has yet to seriously challenge the orthodox economic policy in the country. Wales has not abandoned growth, assumes decoupling is possible, and remains very much wedded to a neoliberal conception of competition. Moreover, the goals of 'prosperity', which we have already considered, and 'global responsibility' are loosely defined. Wales has potentially ground-breaking legislation, but, as yet, it appears unsure how to fully implement it.

The friction of distance

The challenge of being a globally responsible nation is a significant dilemma that places and nations must resolve within the context of sustainable place-making. In our view, a place cannot be sustainable unless it is globally responsible. But what does global responsibility mean? Typically, global responsibility discussions tend to begin in the context of the global climate emergency. It is worth noting here that increasingly nations are signing up to emissions reductions targets, although some appear to have clearer strategies and have made more progress in achieving these targets than others.

One thorny issue relates to how emissions targets are calculated. There are essentially three ways in which government calculates emissions: territorial, production, and consumption-based. The World Wildlife Fund (2020) offered the following definitions:

- Territorial-based emissions: *emissions and removals taking place within national (including administered) territories and offshore areas over which the country has jurisdiction (not including international aviation or shipping).*
- Production-based emissions: *emissions caused by national residents and industry whether within the national boundaries or abroad, but excludes emissions within the national boundaries which can be attributed to overseas residents and businesses.*
- Consumption-based emissions: *emissions allocated to consumers in each country that are trade adjusted, therefore consumption equals production minus exports plus imports.*

The consumption-based method, the WWF suggests, represents a country's true carbon footprint. It shows that, for example, although between 1990 and 2016 the UK reported a 41 percent reduction in greenhouse gas emissions within the country's borders, over the same period its consumption-based emissions declined by just 15 percent. This difference was the result of emissions relating to goods and services imported from overseas. In 1990, the overseas proportion of the UK's carbon footprint was 14 percent, but by 2016 this had risen to 46 percent. (WWF, 2020).

The impact of financial sectors on global emissions is also a major issue. Recent research by Greenpeace and the WWF has shown that emissions associated with the UK's private financial institutions, global activities where investments have been made, are 1.8 times larger than the UK's domestically produced emissions. Indeed, the research estimates that if the UK financial sector were a country, it would be the 8th largest polluting country in the world. (Greenpeace/WWF, 2021)

For us, although it is clearly important for nations to reduce and eliminate their territorial emissions, particularly as these are areas where they have more immediate control, such reductions are only partial. Emissions targets should ultimately be based on consumption. It is not acceptable for richer nations to outsource their emissions through globalised supply chains, nor for their independent financial institutions to invest in carbon-intensive activities elsewhere. Such activities are ultimately self-defeating, as emissions produced in another country are just as damaging to the global environment as those produced within the sovereign territory. This is exemplified by the Australian economic dependence on massive coal and gas imports to China, India, and SE Asia where its contribution to global warming is not considered in the Australian government's continued attempts to obfuscate its carbon reduction targets.

The obvious way to achieve real carbon reductions, of course, is to reduce overall consumption. For those products where consumption is clearly necessary, relocalising production would force governments and places to take more direct control over how those things are produced, and make them more directly accountable. This would also create an opportunity for localised forms of sustainable employment, a theme we return to in the case studies in Part Two. If consumption declines and/or production is relocalised, however, this does raise the question of what happens to places that currently produce such products, many of which are in poorer counties? Adding to this dilemma is the fact that many of the countries that lay great claim to being 'globally responsible' continue to compete for 'investment'. This we may call the 'friction of distance'. Countries find themselves in globalised neoliberal competition to attract inward investment of footloose capital away from other countries, whilst at the same time more and more goods and services are imported largely to increase profit, whilst emissions are correspondingly offshored. This is the economic dilemma that lies at the heart of this book, and why we argue for the need for a new economic model for people and planet.

Economies for sustainable place-making

There have been various attempts to imagine more sustainable economic models. Reed and Lawson (2011) described what they called the 'core economy' – the human resources that comprise and sustain social life – which is '…embedded in the everyday lives of every individual (time, wisdom, experience, energy, knowledge and skills) and in the relationships among them (love, empathy, responsibility, care, reciprocity, teaching and learning)' (p. 32). Robin Murray (2009) suggested the 'social economy' is the location within which issues such as inequality, the environmental crisis, and the ageing society are being addressed. Kate Raworth (2017) has warned 'don't wait for economic growth to reduce inequality – because it won't. instead, create an economy that is distributive by design'. She says, 'if large-scale actors dominate an economic network by squeezing out the number and diversity of small and medium players, the result will be a highly unequal and brittle economy' (pp. 174–176). She concludes that we need diversity and distribution. Each of these commentaries is salutary lessons.

Central to our understanding of the economic model needed for sustainable place-making is that the sustainable economy is the economy of the things we do cooperatively for the global common good, and in ways that are environmentally beneficial rather than environmentally destructive. In the case studies that follow in Part Two, we begin to sketch out what we believe an economic model for sustainable place-making looks like in particular places. For us there are four fundamental components to such a model, which underpin our approach to each case study, and which we shall return to in more depth:

- Distributed economies – *the economic model of wholistic sustainable place-making is rooted within communities, rather than agglomerated dense urban areas. There is an emphasis on the localised economies producing a greater balance of the goods and services that are essential for human and environmental well-being.*
- Non-extractive economies – *the economic model of sustainable place-making produces goods and services in ways that do no harm to local or global environments. For us, it is possible to produce all the goods and services we need without causing environmental degradation.*
- Socially just economies – *the economic model of sustainable place-making delivers social justice and overcomes economic inequalities.*
- Cooperative economies – *the economic model of sustainable place-making is a cooperative and collective rather than competitive economy. As places produce more of what they need locally, they are freed of the current mantra of competing for 'investment'. In Chapter 10, we explore this issue further as we develop the concept of 'regenerative collectivism'.*

Arturo Escobar (2008) explored various place-based differences in the wider context of globalisation. *Buen Vivir* are the words used in Latin America to describe alternatives to development focused on the 'good life'. Eduardo Gudynas (2011) suggested that although the term is difficult to translate into English, its meaning combines classical ideas of 'quality of life', with the understanding that 'well-being is only possible within a community', where the community is understood as both cohabiting with others and nature (p. 441). There have been various approaches to *Buen Vivir*, most notably in Bolivia and Ecuador, and Gudynas suggests the term itself is best understood as an umbrella for a set of different positions. There are, however, a set of ideas that are common to the concept. All positions promote ethical perspectives grounded in values that are a rejection of utilitarianism, which reduces life to an economic value. Instead, intrinsic values are recognised, including those of and in nature. It also rejects growth as the basis for development. These perspectives accord very closely with our own.

In summary

Neoliberalism is hegemonic in economic policy discourse, and growth is a central pillar of neoliberal ideology. For neoliberalism, competition is the defining characteristic of human, and therefore economic, relations. The state only endangers the optimum performance of the free market and should therefore withdraw. The UK and the USA have led the retreat, but in their quest to prove their commitment to the 'end of ideology', they have instead found themselves 'dancing with dogma'. Whilst capital has accumulated under the ensuing conditions, labour has been subjugated, inequality has grown and the poorest in society have been increasingly marginalised. As the rewards of growth are felt by an ever-diminishing elite, and employment is no longer a guaranteed route out of poverty, the central objectives of regional economic development – jobs and growth delivered via the successful competition for inward investment – are increasingly flawed. Against this background, the logic of growth is being challenged as its foundations are crumbling.

Growth as GDP does not measure detrimental social, environmental, or cultural costs, nor does it quantify those things that are central to well-being. In the quest to find alternatives and to reduce the negative impacts of growth, the concept of decoupling has gained significant traction. Born of a well-meaning desire to reduce environmental degradation, decoupling has been adopted with relish by governments. It offers the prospect of overcoming some of the environmental impacts of growth, without fundamentally endangering the existing economic order. As discussed in this chapter, however, there are serious and persuasive reasons to question whether decoupling can ever achieve its goals. Decoupling is a compromise that is unlikely to deliver the ecological shift required, but proponents of change are often divided in their advocacy – degrowth and Green New Deal are two such divisions. Whereas supporters of Green New Deal argue that growth is necessary to fund (largely via increased tax returns) the ecological transition, degrowth supporters argue that a growing economy makes environmental transition more difficult to achieve as the goal posts are constantly moving further away. For us, economic growth is not a prerequisite for ecological modernisation, nor is it necessary to fund the transition.

As GDP growth is not an indicator of well-being, and current models do not deliver social or environmental good, different measurements and systems are necessary. As countries like New Zealand, Bhutan, and Wales, and individual US states like Maryland, are seeking to address the measurement issue, finding alternative models that are consistent is proving more difficult. Wales, for example, now has genuinely pioneering well-being legislation but has also largely pursued neoliberal growth-based and competitive economic policies. Global responsibility is a particularly difficult concept that such countries are finding it difficult to resolve. For us, economies for sustainable place-making need to be designed from the perspective of global good. As we explore in Part Two, economies need to be distributed, non-extractive, socially just, and cooperative, rather than competitive. In the chapters that follow, we explore further the social, environmental, and cultural dimensions of sustainable place-making.

References

Attenborough D (2020) *A Life on Our Planet Earth: My Witness Statement and a Vision for the Future*. London: Witness Books.

Bell D (1960) *The End of Ideology*. London: Harvard University Press.

Bleys B and Whitby A (2015) 'Barriers and opportunities for alternative measures of economic welfare', *Ecological Economics* 117(c): 162–172.

Christophers B (2018) *The New Enclosure: The Appropriation of Public Land in Neoliberal Britain*. London: Verso.

Dorling D (2019) 'The biggest story in the UK is not Brexit. It's life expectancy'. Retrieved 14 January 2020, from https://thecorrespondent.com/177/the-biggest-story-in-the-uk-is-not-brexit-its-life-expectancy/189834797460-1bfa5858

Ellen MacArthur Foundation (2016) *Intelligent Assets: Unlocking the Circular Economy Potential*. Isle of Wight: Ellen MacArthur Foundation.

Engelen E, Sukhdev S and Williams K (2016) 'From Engels to the grounded city', *Wuppertal KW Conference Paper*. Manchester: CRESC.

Escobar A (2008) *Territories of Difference. Place, Movement, Life, Redes*. Durham, NC: Duke University Press.

European Commission (2001) *Environment 2010: Our future, Our Choice – 6th EU Environment Action Programme 2001–2010*. Luxemburg: European Commission.

Fioramonti L (2017) *Growth is dying as the silver bullet for success. Why this may be good thing* (The Conversation, 28/05/17).

Florida R (2005) *The Flight of the Creative Class: The New Global Competition for Talent*. New York: HarperCollins.

Freeman D (2020) *Can Globalisation Succeed?* London: Thames and Hudson.

Fukuyama F (1992) *The End of History and the Last Man Standing*. London: Penguin.

Gilmour I (1992) *Dancing with Dogma Britain under Thatcherism*. London: Simon and Schuster.

Glaeser E (2011) *Triumph of the City*. London: MacMillan.

Gopinath G (2019) 'The World Economy: Synchronized Slowdoen, Precarious Outlook', *IMF Blog*, 15 October 2019: https://blogs.imf.org/2019/10/15/the-world-economy-synchronized-slowdown-precarious-outlook

Greenpeace/WWF (2021) *The Big Smoke: The Global Emissions of the UK Financial Sector*. London: Greenpeace UK/WWF UK.

Gudynas E (2011) 'Buen Vivir: Today's tomorrow', *Development*, 45(4): 44–447.

Gulliver K (2017) *Human City Manifesto: Realising the Potential of Citizens and Communities in the Shared Society*. Birmingham: The Human City Institute.

Hayek F (1993 Edition) *The Road to Serfdom*. London: Routledge.

Hildreth P and Bailey D (2013) 'The economics behind the move to "Localism" in England', *Cambridge Journal of Regions, Economies and Societies*, 6: 233–249.

Jackson T (2009) *Prosperity without Growth: Economics for a Finite Planet*. London: Earthscan.

Jackson T (2021) *Post Growth: Life after Capitalism*. Cambridge: Polity.

JRF (2013) *In-work poverty*, Retrieved 15 June 2021, from https://www.jrf.org.uk/mpse-2015/work-poverty

JRF (2021) *UK Poverty 2020/21*. York: Joseph Rowntree Foundation.

Kallis G, Paulson S, D'Alisa G and Demaria F (2020) *The Case for Degrowth*. Cambridge: Polity.

Krugman P (1998) 'What's new about the new economic geography?', *Oxford Review of Economic Policy*, 14: 7–17.

Kubiszewski I, Costanza R, Franco C, Lawn P, Talberth J, Jackson T and Aylmer C (2013) 'Beyond GDP: measuring and achieving global genuine progress', *Ecological Economics*, 93: 57–68.

Lang M and Marsden T (2018) 'Rethinking growth: towards the well-being economy', *Local Economics*, 33(5): 496–514.

Lang M and Marsden T (2021) 'Territorialising sustainability: de-coupling and the foundational economy in Wales', *Territory, Politics and Governance*. https://doi.org/10.1080/21622671.2021.1941230.

Love M (2017) *Mapping Economic Potential in North East Glasgow: Exploring the Case for Asset Based Community Development*. Glasgow: Common Weal.

Martin R (2008) 'National growth versus spatial equity? A cautionary note on the new "Trade-Off" thinking in regional policy discourse', *Regional Science Policy and Practice*, 1: 1–13.

Mason P (2015) *Post Capitalism: A Guide to Our Future*. London: Allen Lane.

Mastini R, Kallis G and Hickel J (2021) 'A Green New Deal without growth?', *Ecological Economics*, 179: 1–9.

Mathur A (2019) 'Rethinking the Green New Deal: using climate policy to address inequality', *National Tax Journal*, 72(4): 693–722.

McClure T (2021) 'New Zealand moves on from "wellbeing budget" to focus on Covid recovery', *The Guardian 4/5/21*, Retrieved 15 June 2021, from https://www.theguardian.com/world/2021/may/04/new-zealand-moves-on-from-wellbeing-budget-to-focus-on-covid-recovery

Murray R (2009) *Danger and Opportunity: Crisis and the New Social Economy*. London: NESTA.

NEF (2008) *A Green New Deal: Joined-up Policies to Solve the Triple Crunch of the Credit Crisis, Climate Change and High Oil Prices*. London: New Economics Foundation.

OECD (2011) *Towards Green Growth*. Paris: OECD.

O'Hara P (2008) 'Principle of circular and cumulative causation: Fusing Myrdalian and Kaldorian growth and development dynamics', *Journal of Economic Issues*, 42(2): 375–387.

Overman H (2013) 'The economic future of British cities: what should urban policy do?' Retrieved 23 October 2019, from http://spatial-economics.blogspot.co.uk/2013/01/-the-economic-future-of-british-cities_23.html

Parrique T, Barth J, Briens F, Kerschner C, Kraus-Polk A, Kuokkanen A and Spangenberg J H (2019) *Decoupling Debunked: Evidence and Arguments against Green Growth as a Sole Strategy for Sustainability*. Brussels: European Environmental Bureau.

Pike A and Tomaney J (2009) 'The state and uneven development: The governance of economic development in England in the post-devolution UK', *Cambridge Journal of Regions, Economies and Societies*, 2: 13–34.

Piketty T (translated by A Goldhamer) (2014) *Capital in the 21st Century*. Cambridge, MA: Belknap.

Pilling D (2018) *The Growth Delusion*. London: Bloomsbury.

Pollin R (2019) 'Advancing a viable global climate stabilization project: degrowth versus Green New Deal', *Review of Radical Political Economics*, 51(2): 311–319.

Raworth K (2017) *Doughnut Economics: Seven Ways to Think Like a 21st-Century Economist*. London: Random House Business Books.

Reed H and Lawson N (2011) *Plan B: A Good Economy for a Good Society*. London: Compass.

Schenker J (2017) 'Revealed: the insidious creep of pseudo-public space in London'. *The Guardian*. Accessed at: https://www.theguardian.com/cities/2017/jul/24/revealed-pseudo-public-space-pops-london-investigation-map

Solly M (2018) 'U.S. life expectancy drops for third year in a row, reflecting rising drug overdoses suicides'. Retrieved 20 September 2019, from https://www.smithsonian-mag.com/smart-news/us-life-expectancy-drops-third-year-row-reflecting-rising-drug-overdose-suicide-rates-180970942/

Spretnak C and Capra F (1986) *Green Politics: The Global Promise*. London: Paladin Collins.

Stiglitz J (2009) *Report of the Commission of Experts of the President of the United Nations General Assembly on Reforms of the International Monetary and Financial System*. New York: United Nations.

Tyler G (2021) *Briefing Paper: Food Banks in the UK*. London: House of Commons Library.

United Nations EP (2011) *Towards a Green Economy: Pathways to Sustainable Development and Poverty Eradication – A Synthesis for Policy Makers*. New York: United Nations.

Welsh Government (2017) *Prosperity for All: Economic Action Plan*. Cardiff: Welsh Government.

Whittaker J (2020) 'Modern monetary theory: the rise of economists who say huge government debt is not a problem', *The Conversation* 7 July 2020, Retrieved 15 June 2021 from https://theconversation.com/modern-monetary-theory-the-rise-of-economists-who-say-huge-government-debt-is-not-a-problem-141495

World Bank (2012) *Inclusive Green Growth: The Pathway to Sustainable Development*. Washington, DC: World Bank.

WWF (2020) *Carbon Footprint: Exploring the UK's Contribution to Climate Change*. Woking: World Wildlife Fund UK.

Zink T and Geyer R (2017) 'Circular economy rebound', *Journal of Industrial Ecology*, 21: 593–602.

3

PLACE AND SOCIAL STRUCTURE

We began this book with a short overview of the potential influence of place on social and economic outcomes. This chapter explores these issues and more closely examines the relationship between place, social structure, and poverty. It will investigate the patterns of poverty and inequality in a globalised world, to identify the varied experience of a global division of labour and its contemporary relations of exploitation that have their origins in slavery, the industrial revolution, and the era of colonialisation that has so fundamentally shaped the world. Whether experiencing the contemporary world from a colonising nation or from the direct experience of the colonised nations, history casts a long shadow, and we require a review of economic and social processes that have brought us to the world we inhabit today.

Place and social structure

This account begins with a brief review of how the contemporary social structure and spatial distribution of poverty have evolved in the 'developed' nations. This will be achieved by examining patterns of social change in the UK as indicative of the development trajectories throughout Europe. With the increased productive capacity created by the agricultural revolution from the 16th century onwards, population growth derived from additional food supply merged with the wealth derived from an increasingly global economy grounded in slavery, to create the conditions for rapid economic expansion, industrialisation, and urbanisation. Coupled with the innovative technology of steam power and developments in metal production, mining, and transport, the foundations of a transformative industrial society were laid over a 200-year period. By the end of the 19th century, the broad brush of social structure in the capitalist nations

DOI: 10.4324/9781003221555-4

was clearly demarcated. The primary class boundaries identified in the work of early sociologists have been blurred by more complex structures of ownership and control of the productive process, and additional sources of identity have consequently been included in contemporary analysis of social relations. The influence of the social class boundaries established during the period of 'organised capitalism' (Lash and Urry, 1987), however, remains significant in determining the patterns of poverty and social exclusion evident in contemporary European society.

The post-war settlement

Our account of the relationship between poverty and place will begin in the UK with the post-World War Two Settlement, which saw a significant break with the past and the emergence of a new paradigm to condition the relationship between citizens and the state. It commences with the seismic shift in the economy, culture, and politics of the immediate post-war period. The period after 1945 saw a fundamental restructuring of the role of government. Ushering in the period of what has now become known as 'big government', the development of a range of enhanced public services such as health, housing, education, and transport saw the government taking responsibility for the well-being of all citizens. Core principles such as 'universalism' saw the birth of 'welfare rights' that each citizen could expect to be fulfilled by government at times of crisis or negative change in their lives. Unemployment and sickness benefits provided income during periods of loss of livelihoods and introduced a financial security for every section of the population. In contrast to the deprivations of the pre-war 1930s, the post-war period saw the development of the welfare safety net that became the norm until the mid-1980s.

Following their landslide victory in 1945, the Labour government was the primary architect of the welfare state. Capturing the post-war social mood for change and reward for the struggle against fascism, public opinion carried all political parties in a similar direction to establish a consensus in support of the welfare state. Ironically, the Labour government achieved its reforms by implementing the proposals of liberal thinker Sir William Beveridge in the famous Beveridge Report (officially called Social Insurance and Allied Services, cmd 6404) published in 1942. The post-1945 period of welfare development is often referred to as the 'golden age of the welfare state' (Wincott, 2013), which lasts into the mid-1970s. Whilst the perception of a golden age may have a degree of mythology attached to it, there is a clearly recognisable period of continued development of social protection measures in the UK and throughout Europe. For Piketty (2013), this was a brief interlude in rising inequality that had endured for two centuries and illustrates the impact that progressive taxation and redistributive policies can have in creating a more equal society (Savage, 2014).

The 1950s and 1960s saw a period of unprecedented development of welfare services and associated welfare rights that addressed the 'five giants', identified

by Beveridge as: Want (poverty), Disease (ill-health), Ignorance (lack of educa-
tion), Squalor (poor housing), and Idleness (unemployment) (Beveridge, 1942).
In turn, the Education Act 1944, the National Insurance Act 1946, the National
Health Act 1946, and the National Assistance Act 1948 addressed these issues and
laid the foundation for a political consensus that survived transitions between
Labour and Conservative governments until the early 1980s.

Closely associated with this period is the adoption by all UK governments
of the Keynesian economic strategy of demand management by a combination
of monetary and fiscal policy. Fine tuning of money supply, taxation levels, and
government expenditure was believed to have the capacity to iron out the peaks
and troughs of the economic cycle and avoid the boom-and-bust economies that
capitalism had historically experienced. The period is often characterised by the
concept of 'corporatism', a coalition between government, business, and organised labour. A version of the wider body of the political theory of pluralism, it
was felt that government, the state institutions, and economic policy could serve
the interests of all of society and balance sectional interests, by producing benefits
for all. Successive governments pursued economic stability and full employment
as their primary purpose. A key objective of this period of welfare capitalism was
also the redistribution of wealth through progressive taxation, where high taxation of high-income earners and businesses supported welfare payments for the
less fortunate.

The sociological analysis focused on the consumer society and the rise of the
'affluent worker' (Goldthorpe et al., 1969) associated with manufacturing and the
productive and extractive industries. Employment in nationalised industries, such
as coal extraction and steel manufacture, created what was seen as a 'labour aris-
tocracy' earning good wages in lifetime jobs and with generous retirement pensions. Workers in manufacturing, transport, the health service, and the education
system experienced employment conditions and reward levels unprecedented in
the period before the Second World War. The growth of a large 'middle class'
was reflected in the suburbanisation of major cities, and for those in the lower
reaches of the labour market, the building of council houses provided safe and
secure housing for life.

The apparent golden age, therefore, created new social boundaries and patterns
of consumption. The urban spaces occupied by industrial workers and those pop-
ulating the bureaucratic systems of the burgeoning welfare state became places of
conspicuous consumption with rising home and car ownership, the emergence of
foreign holidays, and the purchase of consumer durables for the home. The period
lay the foundations of the future spatial inequalities that would develop in the
1980s, as these employment patterns collapsed, and what became post-industrial
communities developed high rates of poverty and social exclusion. Even in this
period of apparent universal affluence, however, there were marginalised groups
in society. The pioneering work of Abel Smith and Townsend (1965) represented
an emerging academic interest in poverty. Their work remains an important contribution, despite recent caveats concerning their findings (Gazeley et al., 2017).

They identified rising poverty despite improved welfare provision, particularly for single-person households. They estimated 2.25 million children were living below the poverty line in 1960. It took a 1966 BBC drama, Cathy Come Home (Fitzpatrick and Pawson, 2016), to alert public consciousness to the social issues, as the overriding perception of the period remained one of social progress and rising affluence.

A full examination of this period would require a book in itself to fully explore the expansion of the state into consumer rights, slum clearance, public housing, infrastructural development, and sponsorship of culture and the arts. One of the primary reasons for this exploration, aside from understanding the period as a prelude to what follows, is to illustrate that the neoliberal economy and its political structures are not the only form that capitalism can take. The period briefly explored above was one in which governments pursued full employment, a redistribution of wealth, and improved citizen well-being. Contrary to the 'there is no alternative' claims of Margaret Thatcher, examination of the immediate post-war period illustrates that capitalism is variable and capable of reform to ameliorate the worst impacts of the free market. Its socio-economic relations can be redesigned to achieve objectives other than the permanent growth and the concentrated wealth accumulation that has characterised neoliberalism. Ironically, 'the golden age' of welfare also saw economic growth and the rise of a consumer culture that, through our contemporary lens, we can identify as the cultural foundations of much of the environmental crisis.

The neoliberal counter revolution

Just as the post-war welfare consensus had been ushered in by two key thought leaders, Beveridge and Keynes, so the challenge to it and subsequent transition to neoliberalism was triggered by the work of two further thinkers, Friedrich A. Hayek and Milton Friedman referred to in Chapter 2. Their writings provided a combination of social and economic thought that was, in large part, a reaction against the welfare capitalism described above. The extended economic crisis of the 1970s provided fertile ground for Hayek and Friedman's ideas to influence economic theory in particular. Led by the 'oil shock' of 1972, the emergence of both economic stagnation and inflation, encapsulated in the term stagflation, and the industrial unrest of the mid to late 1970s came to characterise the decade. The corporatist bargain was eroded by successive governments, including the 1976–1979 Labour government of James Callaghan. In this emerging political and economic crisis, Hayek provided both economic and political critiques of welfare capitalism.

Arguing against the economic management fostered by Keynesianism, Hayek asserted the primacy of the market and the impossibility of central planning of economic activity. Clearly directed at Soviet central planning, there were also implications for the nationalisation of key industries in the UK and the social engineering of the welfare economy. Hayek's view was that markets and prices are self-regulated through the multiplicity of individual consumer decisions. This contrasted with conscious planning, which, he argued, could never possess the required level of data and knowledge to provide a stable economic system. Echoes of this view are replete in the celebration of the individual consumer, shareowner, and homeowner model of democracy advanced by Margaret Thatcher and the New Right. Hayek also developed a sophisticated critique of welfarism in his 1944 work the Road to Serfdom, a critique of socialism that also aimed at the social-democratic overlap with his experience of national socialism in 1930s Germany. He saw the economic and social management of welfare capitalism as a 'slippery slope to totalitarianism', because, he argued, the state required ever greater economic control and data on its citizens. Hayek's arguments were adopted by those who already saw the economics of welfare as unsustainable, as they forged a critique of both welfarism and Keynesian economics.

To these powerful arguments, which resonated so strongly with the economic and social conditions of the 1970s, we add the economic theory of monetarism, most cogently presented in the work of Milton Friedman (Cole Julio, 2007; Boettke and Smith, 2016). Friedman challenged Keynesianism on the basis that it presented an unacceptably high risk of inflation, as significant fluctuations in the money supply, driven by political conditions, impacted the economy negatively. Friedman therefore advocated strict control of the money supply with fixed, minor increases determined by a tight regulatory process. He moved from a faith in central banks to achieve this process, to the identification of computer-driven increases in money supply (Boettke and Smith, 2016). Embedded in a wider free-market philosophy, his role as adviser to governments including Ronald Reagan and Margaret Thatcher ensured that a focus on the budget deficit and surplus became one of the central tenets of neoliberalism (Kelton, 2020).

Poverty and social exclusion in industrial and post-industrial society

This brief exploration of the post-war settlement and its subsequent demise provide the context for the emergence during the 1980s of what has been termed the 'new poverty'. The election of a Conservative government in 1979 heralded a dramatic shift in economic and social policy, which is explored later in this chapter. Processes set in motion by the neoliberal politics and economics of the 1980s continue to resonate in the UK, and their impact was compounded by the aftershock of the 2008 Global Financial crisis. The economic policies of the 1979 Conservative government were, as we have seen, informed by the work of Hayek and Friedman. A mantra of ever-increasing competition was perceived as a solution to the low productivity of the British economy and especially the nationalised industries (Bowman et al., 2014). The government embarked on a range of policies that severely limited its economic management functions and introduced market-based processes to a wide range of economic and social spheres. Consequently, there was a rebalancing of power towards capital and a diminution of the bargaining power of trade unions, as well as a reduction of their ability to collaborate with government. Several key processes have been identified by Bowman et al. (2014): trade union reform, dismantling the command economy, and the growth of outsourcing. We will summarise each of these processes below.

First, labour market flexibilisation was achieved by 'trade union reform', which, euphemistically, represented a process of rapid erosion of trade union rights, and the ability of trade unions to organise memberships and exercise the right to strike. The changes of the early to mid-1980s consequently laid the foundation for the low wages, reduced working conditions and diminished employment rights, which enabled employment practices such as zero-hour contracts to emerge during the 1990s. Many of the trade union reforms supported the casualisation of labour, exemplified by the demise of the 'job for life' in nationalised utilities. In manufacturing, the emergence of managerial practices such as 'just in time' production, created a demand for labour flexibility, itself a euphemism for temporary employment contracts without holiday or sickness pay. Job security was sacrificed so that manufacturing capability was able to respond rapidly to market fluctuations in demand.

Second, a programme to dismantle the 'command economy' (Bowman et al., 2014), through the privatisation of state enterprises, eroded working conditions in key economic sectors including transport, energy, and telecommunications. Developing the concept of a share-owning democracy, the Thatcher government privatised utility companies, the railways, and even introduced privatisation into the road systems. Privatisation of mutual financial institutions, such as the building societies, also eroded the sense of collectivism that had informed the establishment and foundation of those institutions. Working conditions and wages were reduced in many of the newly privatised companies, adding to the gradual rise of in-work poverty, which by 2010–2011 replaced welfare dependency as the primary cause of poverty in the UK (McBride et al., 2018). In many instances, workers made redundant upon privatisation were rehired on reduced wages in the weeks that followed. The claims made about the potential benefits to the consumer of private enterprise – price reduction through competition and higher quality services – rarely materialised. Indeed, the cost of utilities actually rose, and the higher energy and water costs for households contributed to the increasing incidence of fuel poverty, water disconnections, and household utility debt.

Third, changes in managerial and employment practices achieved by the growth of 'outsourcing' practices in the public sector, radically reduced professional control of public services. Coupled with a growing audit culture in health and education, these changes brought public services under greater degrees of central government control (Bowman et al., 2014). Where outright privatisation of institutions wasn't possible, for example in health or administrative functions of government, outsourcing services to a growing sector of private providers became the norm. From cleaning services to IT provision, from legal to policy development, the role, scope, and scale of outsourcing companies developed steadily. As in the privatised industries, those employed to deliver the services often did so on reduced wages and conditions. Outsourcing, particularly in manufacturing, became common within a global division of labour, which saw major British industries transfer production to Eastern Europe, India, Southeast Asia, and China to take advantage of the reduced labour costs that could be gained. The loss of jobs in those places previously used to relatively stable employment and high wages had a devasting impact on working-class communities throughout the UK, particularly in Wales, Scotland, and northern England. Government policy did little to halt the industrial decline, favouring financial services over manufacturing. Consequently, major industries such as car manufacturing, steel production, and heavy engineering were significantly reduced in scale.

Understanding the evolution of poverty post-1979

The major economic changes outlined above laid many of the foundations for high levels of unemployment and low wages in post-industrial communities and privatised and deregulated economic sectors. Comparing poverty rates over time is 'notoriously difficult' (Glasmeier et al., 2008, p. 2) and confronts multiple changes in definition and measurement, along with changing cultural definitions of a minimum living standard. Given the general agreement that the 'golden age' of welfare ended in 1979, it is useful to note that the poverty rate at that time (measured as those existing in households with less than 60 percent of median income) stood at 13.4 percent. This had risen to 25 percent by the early 1990s.

In 2018, the UN Special Rapporteur on Poverty and Human Rights, Philip Alston, concluded that Britain was in breach of several human rights for women, children, and people experiencing disability. He also identified a conscious bias on the part of the UK government not to resolve these issues (Alston, 2018). With one-fifth of the UK population (14 million people) living in poverty, and a child poverty rate of 30 percent, he left no doubt about his belief that UK austerity policies since 2010 directly impacted the growth of poverty experienced by these groups. More recently, the Institute for Fiscal studies (IFS) identified a 3 percent growth in household incomes between 2017–2018 and 2019–2020. Largely derived from a growth in employment, it nevertheless found that the lowest income decile had remained stagnant from 2013 to 2014. The IFS analysis also concluded that the relative poverty rate (measured as 60 percent of income in the year of measurement) was 22 percent, and that child poverty stood at 31 percent (Cribb et al., 2021). The impact of poverty on children is pervasive (Lyndon, 2019), and this strongly impacts the lifelong patterns of social exclusion considered below. Our understanding of poverty, developed here, is less concerned with statistical trends, and instead focuses on the lived experience of poverty, as well as its social, cultural, and economic implications for social exclusion.

Contrary to popular belief, the Conservative governments of 1979–1997 did not significantly 'roll back' the welfare state. Public social expenditure in 1997 amounted to 25.1 percent of GDP, a little higher than the level in 1979 (Hills, 2011). Nevertheless, the economic, industrial, and fiscal policies identified above had a significantly corrosive effect on working-class communities throughout the UK. Whilst this was at first felt in the mining and steel-making communities of the UK, awareness of challenging social conditions in British cities was growing by the mid-1980s. The 1981 Scarman Report on the Brixton Riots (Scarman, 1986), together with the 1985 Faith in the City: The Report of the Archbishop of Canterbury's Commission on Urban Priority Areas (1985), drew attention to the rising unemployment and poverty evident in many urban communities. In subsequent policy responses, we see the beginnings of the stigmatisation of the poor as an 'underclass', a modern version of the 'undeserving' poor distinction that had been encapsulated in the Poor Law reforms of the 19th century. By 1990, this was given full voice in the work of Richard Murray (Murray, 1990), which had been forged in the USA and enthusiastically adopted by the UK Conservative Party. According to this, poverty was explained by individual choices to adopt negative lifestyles, characterised by lone parenthood, elective unemployment, criminal activity, and substance misuse. A caricature of the 'the poor' emerged, which ignored structural unemployment in the post-industrial communities and fixed the blame for poverty on the individual choices and failings of the poor themselves.

During this period a deep connection between place and poverty was forged. An early identification of the 'North-South divide' (Philo, 1995), highlighted the contrasting economic performance of South East England and the old industrial communities of a northern Britain. 'The North' quickly became a metaphor for post-industrial Britain. Large peripheral council estates, often constructed to house workers in single enterprise local economies, became the location of patterns of poverty that were increasingly analysed in terms of multiple and cumulative disadvantages. The spatial patterns of unemployment have had a continuing corrosive impact. Clasen (2002, p. 60) argued, 'de-industrialisation is a major cause for the spatial distribution of unemployment and employment which... exacerbates problems of tackling poverty, inequality and social exclusion...'.

Until the 1990s, poverty was generally measured against the concept of a poverty line. Defined differently at various points in the policy cycle, the identification of an income level below which subsistence became difficult, focused analysis on the lack of financial resources available to individuals and families. Consequently, policies to address poverty were based on redistributive measures with support for low incomes through a range of social security benefits. As poverty levels have increased, and we develop further our understanding of the lived experience of poverty, attention has focused on the multiple dimensions of poverty, rather than the simple lack of financial resource. By the mid-1990s, the term 'social exclusion' was increasingly used to explain how a lack of basic income also shapes outcomes for housing, health, and education. These areas of focused disadvantage were identified as the 'triangle of poverty' in work conducted by the Regional Research Programme at the University of Glamorgan (Adamson, 2008). Following the 1991 Census, the research identified major shifts in socio-economic patterns, largely centred on local authority housing estates, where unemployment levels of over 50 percent were identified (Campbell,

1993; Adamson, 2008). These emerging patterns of multi-dimensional poverty were, therefore, increasingly conceptualised as a process of social exclusion (Levitas, 1998).

In the absence of employment and adequate income, alternative patterns of transition to adulthood emerged, which were determined by serial rather than permanent relationships, and were accompanied by high levels of benefit dependency and economic inactivity. The lived experience of poverty often took place within the tight socio-economic boundaries of local authority housing estates, with little external engagement with work, culture, and politics. These patterns effectively prevented participation as fully engaged citizens in the general life of the community. Following this conceptualisation of social exclusion, poverty is now seen as a failure of the state to address the needs of a section of its population, rather than the failings of individuals. Poverty, in this context, is seen as 'relational' rather than 'distributive' in the conventional understanding that prevailed previously (Room, 1995). Accordingly, policy responses focused less on redistribution of income, and more on programmes designed to foster social inclusion to ameliorate the multi-dimensional impact of resource-based poverty.

These policies were exemplified by the Strategy for Neighbourhood Renewal in England, with similar programmes such as Communities First in Wales and Community Scotland (Adamson, 2010; Shaw and Robinson, 2010). A product of the Social Exclusion Unit established in 1997 by the first Blair government, the connection between place and poverty was clearly identified in the 18 Policy Action team reports that informed the design of the Strategy. The devolved policies each drew a direct link between place and poverty and targeted the most disadvantaged communities with programmes that focused on labour market inclusion and community engagement. Building on earlier urban renewal initiatives, these community-led regeneration programmes saw the participation of communities and their members as a necessary element of the solutions that could bring some amelioration of rising poverty levels (Ball and Maginn, 2005). These spatial strategies stood alongside the many variants of the labour market inclusion programmes within the New Deal programme of the Labour government.

For some critics, these policies were symptomatic of the Labour's conversion to neoliberalism, in that they appeared to shift responsibility for solutions to poverty to those experiencing it. The approach also introduced 'workfare' concepts into benefit entitlement (Deeming, 2015), shifting from the income protection perspective of the post-war settlement to a conditionality in benefit qualification regimes. This development of a perception of both rights and obligation in welfare entitlements can be seen as symptomatic of this apparent shift in conventional Labour attitudes to poverty and indicative of Labour's pursuit of a 'third way', lying between the extremes of 'new right' market capitalism and the Keynesian social welfare state (Giddens, 1998). The neighbourhood renewal programmes, however, developed community capacity and community empowerment and did much to change the experience of poverty in the target communities. They were less successful in challenging the structural origins of poverty (Crisp et al., 2015).

Consequently, the period 1997–2010 saw unprecedented levels of expenditure in some of the most deprived communities in the UK. Regrettably, despite this investment, the same communities remain marked by a direct relationship between poverty and place in the 2020s.

Critics of place-based poverty strategies had always maintained that such approaches ignored the distribution of poverty and the fact that most people defined as financially poor lived, and continue to live, in communities that are not defined as disadvantaged. Area-based policies, it is argued, do not address the many poor and excluded households and individuals living outside deprived neighbourhoods, excluding them from any benefits from these policies. More seriously, these area-based initiatives deflect our attention away from thinking about the causes of the problems and their potential solutions as lying outside the deprived areas (Oatley, 2000, p. 89). These arguments promote, as an alternative, the development of people-based policies targeting categories of the poor rather than poor locations. These arguments notwithstanding, the impact of place-based poverty creates localised and compounded patterns that heavily impact social and economic outcomes. Poverty is not simply a result of personal attributes, such as poor education and low skills, it is compounded by the social and economic conditions of places (Glasmeier et al., 2008, p. 5). Environmental factors, cultures of household mobility, and deficits of social capital have each been deployed as explanations for the ways in which place-based social conditions compound the impact of the basic lack of financial resources (Jordan, 2008).

We argue that the term 'social exclusion' captures the experience of whole communities that do not enjoy the benefits of contemporary culture and society. These communities are excluded from economic, social, and cultural outcomes that most of the population achieves. The emergence of patterns of social exclusion was analysed in Wales as a development in the social structure of a marginalised working class (Adamson, 2001). With its origins in the traditional working-class occupations and industries that brought relative levels of affluence, the lived experience of this class fraction was conditioned by the new patterns of poverty following the collapse of the employment that had sustained such communities (Adamson, 1998). In the wider UK context, the Great British Class Survey (Savage et al., 2013) traced these processes and identified a similar cohort referred to as the 'precariat', a term borrowed from Standing (2011). With a household income at that time of just £8,000 and negligible savings, the precariat had very little social, cultural, or economic capital. Savage et al. (2013) also identified the precariat's primary location as the post-industrial areas of the UK, and as a cohort populated by the 'the unemployed, van drivers, cleaners, carpenters, care workers, cashiers and postal workers' (p. 243). A tipping point in the distribution and causation of poverty was reached in 2011 when, for the first time, the majority of those defined as living in poverty could be found in working families, so-called 'in work poverty', rather than the unemployed or those dependent on state benefits or pensions (MacInnes et al., 2013; McBride et al., 2018).

Contemporary poverty policy

This outline of the patterns of social structure, poverty, and place brings us to the current period and arguably the highest levels of poverty in the UK. The UK legislative commitment to end child poverty by 2020, which had been enshrined in the 2010 Child Poverty Act, was repealed by the 2016 Welfare Reform and Work Act. During this period there has also been a shift from measuring child poverty as an income-based measure (60 percent of mean income), to a limited attempt to address multi-dimensional poverty with two 'life chances' indicators: proportion of children in workless households, and educational achievement at age 16. The first of these indicators fails to capture 'in work' poverty, the majority identified in previous income-related measures, and the second is not a measure of poverty at all, rather an indication of future risk of poverty (Dickerson and Popli, 2018). These changes of the definition and measurement of poverty also took place against the backdrop of the post-2008 austerity programme. The measures to reduce the role of big government and the budget deficit saw the ending of place-based interventions. Furthermore, significant welfare reforms saw a reduction in incomes. The roll-out of Universal Credit, the benefits cap, child credit limits to two children, benefit qualification waiting periods, and tighter eligibility requirements have all contributed to reduced incomes for many UK families. These changes have been seen as driving the increasing food poverty evident in the UK and the consequent necessity for the growth in food bank provision (Taylor-Robinson et al., 2013; Möller, 2021).

Ideologically, the Cameron government's approach was underpinned by a concept of 'social justice', which was based on the work of Iain Duncan Smith and the Centre for Social Justice. Crossley (2017) explored the use of this term in government documents and concludes that it departs from accepted previous interpretations. Reviewing the understanding of social justice, from the work of Rawls to its uses in New Labour, Crossley concludes that its contemporary use '...is inconsistent with other theories and understandings of the concept' (p. 21). Instead, the attention is on specific social groups including 'problem families' and those experiencing addiction. It loses any 'focus on structural and institutional issues' (p. 26), localises interventions, and minimises the role of the state, replacing it with charities, philanthropy, and the general civic processes of the Big Society (MacLeavy, 2011). This perspective returns to notions of the 'undeserving poor' and further stigmatises poverty as the consequence of personal failure or a consciously chosen lifestyle.

Finally, mention must be made of the Johnson government's Levelling Up programme. The programme identifies an area-based approach centred on 'ex-industrial areas, deprived towns and coastal communities' (HM Government, 2021, p. 1). With a core focus on local infrastructural renewal and return of 'pride' to communities, the initial stage investments will focus on transport, town centres, and cultural investment (UK Government, 2021). At the time of writing, it is not yet apparent how this will develop local economies and is even less clear how it will impact the deep poverty that we have identified in the preceding discussion.

Poverty and the global south

The benefits of the 'golden age' of the welfare state were, of course, never felt in the Southern Hemisphere (Breman et al., 2019). There, centuries of evolving relationships of exploitation from slavery to the contemporary trade relations of globalisation, have constructed a geo-spatial distribution of wealth and poverty that continues to advantage the colonising nations. In this brief exploration of the continuing impact of that legacy we see that, just as in the example of the UK, global poverty is structural, spatially organised, and maintained by political and economic ideologies that have dominated the internal politics of the colonising nations.

Slavery, colonialism, and post-colonialism

Any account of global uneven development, and its consequence on the global distribution of poverty, must begin with a consideration of the impact of slavery, particularly on the African continent and, indeed, on Britain itself. Often presented as an over-simplified 'triangle of trade' model (Campbell, 2021), which limits the analysis to the cotton, tobacco, and sugar trade routes, the reality is that enslavement was a core component of the entire northern hemisphere economy. The economic surplus derived from slavery underpinned the social and economic development of the European nations, with the UK at its forefront. It is impossible to explain the burgeoning wealth of the British mercantile era and its substantial funding of the industrial revolution without recognising that it was founded on wealth derived from enslavement. With £20 million paid as compensation to slave owners following emancipation in 1834, this alone contributed significantly to the wealth of certain families and elites in the UK. That many of them had benefitted directly from the slave trade is evidenced in their compensation claims (Hall, 2020), and it highlights the concentrations of wealth that were accrued. Furthermore, benefits were not limited to the national elites but percolated to every level of society (Donington et al., 2022). Fundamentally, the methods and relations of production evidenced within the plantations of the British West Indies represented early forms of capitalism, which were yet to develop in the domestic British economy as it slowly transitioned from feudalism to capitalism (Armstrong, 2019). These factors made enslavement fundamental to the development of European capitalism.

Whilst the slave trade fuelled this economic development process in Britain and Europe, it had the opposite impact on the enslaved nations. The destruction of indigenous models of social organisation, cultures, and economic trading laid the foundations for the impoverishment of colonised nations and their people. The primary impact was a demographic distortion of much of sub-Saharan Africa, with the forced removal of some 12 million people who from 1500 onwards were sent as slaves to the New World. 1.5 million of these enslaved people perished in the 'middle passage' en route to the Americas. Often overlooked are the 6 million people who were sent as slaves to the 'orient' and the 8 million enslaved people within Africa itself, practices that continued long after European abolition (Inikari and Engerman, 1992). The secondary economic impacts of this human trade were also significant, with some economic benefits accruing in the West African coastal centres of the slave trade, as well as to the raiders and traders who directly profited. Less visible is the impact on the trajectory of African economic development, as the distorted trade patterns prevented the development of market institutions and an African response to the increased trading opportunities of expanding European nations.

As well as the financial impact of the slave trade, its impact on cultural perceptions of Africa, its populations and of later colonies was considerable, as theories of racial superiority of the European peoples grew:

> Assumptions about the laboring capacities of those with dark skin, fears of pollution and degeneration associated with mixed populations, or the conviction of White capacity to rule over others: these were ideas that circulated in the metropolitan world and infiltrated the practices of everyday life as they were reconfigured in imperial Britain.
>
> *(Hall, 2020, p. 175)*

Later characterised by Edward Said as 'orientalism' (Said, 1978), ideologies of racial and cultural supremacy sought to justify and rationalise the legacy of slavery and the continued exploitative relationships of colonialism and post-colonialism.

The economic and social relations of colonialism evolved over an extended period and the outcomes were path-dependent, incorporating local elites in different ways to serve the purposes of the colonial power. Early roles for the mercantile trading companies, effectively acting as agencies of the state, were replaced by direct rule, a process exemplified in India with the transition to the British Raj from the East India Trading Company. This period of direct colonisation paradoxically saw considerable social and infrastructural development, but often at the expense of prior indigenous commerce and manufacturing, the textile industry in the case of India. The model of an oppressing power and a subjected people disguises complex webs of social relationships that include acts of genocide, resistances, cultural adaptations and mergings, and reciprocity and complicity in and benefits by indigenous elites in the economic systems of the colonial power structure. In simple geo-political terms, the era of colonialism is considered to have ended with the independence movements of the 1950s and 1960s and the notional freedom of nations ceded at the time by the colonial powers. In reality, of course, the constitutional and legal structures of the independent nations, their continued economic relationships, and their cultural formations extended the primary characteristics of colonialism that can be defined as an 'ongoing' process (Cipolla and Howlett-Hayes, 2015).

The long-term spatial consequences of the division of Africa, Asia, and Latin America into artificial nations, which reflected the agreements and conflict-driven negotiations of the colonising European powers, are clearly visible. These artificial divisions, which ignored historical economic and cultural relationships between peoples, have done much to demarcate the boundaries of poverty in the global community. One of the legacies of the colonial period is the stunted economic development of many of the nation states that emerged when colonial powers withdrew. Conceptualised historically as a distinction between the global North and South, one of the primary spatial divisions is that between a rich and developed Northern Hemisphere and a poor and undeveloped Southern Hemisphere. Within poor nations, the role of place in determining social structure has primarily been a feature of rapid urbanisation and the growth of informal settlements. Whether in the favelas of Latin America or the slums of India, we see concentrations of populations drawn from rural histories into the modernity of the city by employment opportunities that are not matched by the provision of housing and ancillary services. The makeshift responses have become locations characterised by both human ingenuity and concentrated poverty.

Globalisation

Whilst the relationships established in the colonial era remain hugely significant, there is now a tendency to see this fractured structure in terms of globalisation. Slavery and colonialism were specific phases of a globalisation process (Freeman, 2020), which have, more recently, been projected as one characterised by free trade and universally beneficial, global systems of production and consumption. These perceptions of the global economy are embedded in the neoliberal celebration of the 'end of history' (Fukuyama, 1992), heralded by the collapse of the Soviet Union and the perceived global ascendency of liberal democracy and capitalism. Whilst Fukuyama's work has undoubtedly been over-interpreted, the hegemonic neoliberal view is that capitalism and free trade had proved their superiority and that a global trading system, structured by multi-national trade agreements, can bring wealth to all nations in a global version of the 'trickle down' theory' that informed social policy in much of Europe. This accelerated a global division of labour that shifted production to less-developed nations at lower cost, with the consequences for the now post-industrial regions of Europe that we explored in the first part of this chapter.

For Freeman (2020), the most recent phase of globalisation, driven by neoliberalism, creates an 'open polity trilemma' (p. 31) in which achieving a balance between global capital flows, democracy, and the role of the sovereign state becomes impossible. In the increasingly globalised economy attracting international capital creates competition between states to maximise the return to capital. As discussed in Chapter 2, reductions in corporate taxation, investment incentives, and deregulation became the tools to compete in the global economy and attract Foreign Direct Investment. For Freeman, this economic policy comes at the expense of democracy, as states are required to reduce welfare budgets to underpin the loss of revenue and deregulation of the economy that is necessary to compete. Monbiot similarly sees the modern neoliberal state as requiring, 'freedom from the demands of social justice, from environmental restraints, from collective bargaining and from taxation that funds public services' (Monbiot, 2016, p. 4).

Similar requirements are placed on 'developing' nations by key international agencies including the IMF, World Bank, and, in pursuit of austerity, the European Union. The consequences are continued poverty in emerging economies as the benefits accrue to transnational corporations and national elites, rather than resolving historical levels of inequality and poverty. This 'hyper-capitalism' (Piketty, 2021) has maintained income inequality within and between nations. From a review of 173 countries, the World Inequality Report concludes that the share of the poorest 50 percent of the population within nations can vary between 5 percent and 25 percent of total income. Even more marked is the distribution of wealth where the poorest 50 percent of the global population own less than 2 percent of wealth, in contrast to the richest 10 percent with a 76 percent share of wealth (Chancel et al., 2022).

These figures financialise poverty and see it as a simple condition of income levels. Using the United Nations-defined minimum income of US$1.90 per day, some 10 percent of the global population (734 million people) live in poverty. As in our analysis of poverty in the UK, however, we recognise that poverty is multi-dimensional. The Multi-Dimensional Poverty Index (UNDP and OPHDI, 2020) addresses this issue by measuring a range of indicators in health, education, and standard of living. These include indicators of nutrition, child mortality, utility access, and housing. This creates a comprehensive view of global poverty that, in the Multi-Dimensional Poverty Index 2020 report, identifies a major reduction in poverty in 65 countries – 4 of which halved the poverty rate since the launch report of 2010 and a further 10 almost halving it. In 14 Sub-Saharan nations, it increased, however, and nearly 85 percent of those in continued poverty live in either that region or Southern Asia (UNDP and OPHDI, 2020).

These conclusions, and those based on studies of the Gini coefficient, have prompted a debate about convergence between wealthy and poor nations. The Gini coefficient, where zero represents perfect equality and one represents absolute inequality, has been widely used as a standard measure of inequality. Emerging as a 'counter-narrative' (Hickel, 2017) to the work of Thomas Piketty (Piketty, 2013) and the Oxfam global wealth distribution reports, Milanovic (2016), by focusing on income inequality rather than wealth inequality, has promoted the thesis of narrowing income differentials or convergence. These claims were quickly deployed to justify neoliberalism, free trade, and 'free-market capitalism' (Hickel, 2017). Critics of this approach, however, have argued that if the rapid income growth of China is removed from the calculation, global income inequality has barely changed (Bangura, 2019). Problems have also been raised regarding the Gini coefficient itself, as well as Milanovic's application of it to a single global calculation as if all humanity lived in one nation (Hickel, 2017).

Hickel (2017) revised calculations of GDP per capita to show that when comparing nations divergence rather than convergence has occurred, with the richest nations, significantly outpacing the income growth in poorer nations. Indeed, Hickel showed that the income ratio between the richest and poorest countries rose from 55:1 in 1980 to 134:1 in 2000, a period of rapid globalisation (although it was noted that there was a slight fall to 118:1 by 2010). Bangura (2019) made similar observations about per capita income. Whilst recognising convergence in key aspects of human development, including health and education, Bangura identified large gaps in social protection in poorer nations that drive global migration trends. Lack of social protection effectively means precarious populations have no route to the survival of loss of employment or subsistence income sources. Bangura argued for a poverty line of US$5 per day, which would identify 4 billion people living in poverty. They also found no clear evidence that convergence would continue at its current pace, and that cyclical economic patterns, particularly in Africa and Latin America, are likely to reduce the pace of convergence. Given that the conclusions reached in the literature we have considered here are dependent on the methods used, these debates have become somewhat arcane and inconclusive. For us, the reality is likely to lie somewhere between the extremes. Whilst there have been considerable improvements in the technical measures of global poverty, the quality of life for many of the global poor remains unimproved, but there is some hope that change could occur.

Indigenous poverty and legacies of cultural trauma and genocide

Another aspect of global poverty that cannot be ignored is that experienced by First Nations people. Perhaps without exception, indigenous people in colonised nations have historically experienced frontier wars, massacre, population decimation, land dispossession, exile to missions and reservations, enforced cultural assimilation of children, language extinction, and generations of cultural trauma. Whether in Canada, the USA, South America, or Australia, racial oppression was historically normalised to the extent that it remains embedded in contemporary relations between First Nations people and what are now their host nation states. In Australia, there were 300 documented massacres with 8,000 aboriginal deaths and 170 settler deaths. 'Dispossession, violence and the introduction of new diseases' reduced populations by between 30 percent and 80 percent by 1900 (Commonwealth of Australia, 2020a, p. 1.4). What amounted to slave labour occurred on farms and cattle stations in the 'stolen wages' of indigenous workers (Gray, 2007). The relocation of aboriginal people to missions and reservations removed people from their Country and Kin, a core element of their identity and central to their culture. There, they were denied the use of their language and the names given to them by their community. Most controversial has been the 'stolen generations' of aboriginal children removed from their families in a crude attempt at cultural assimilation. The policy continued between 1910 and

1972, and it is estimated up to a third of children were removed from institutions or adoption by white families. Similar policies were enacted in Canada and the USA (Douglas and Walsh, 2013).

The cultural trauma of this experience continues to resonate through Australian aboriginal communities today as a direct legacy of colonisation. In 2018–2019, 27,200 people over 50 directly experienced first-hand what it was like to be part of a stolen generation. Analysis of the impact on children living in households with a member of the stolen generations (7,900 in 2018–2019) shows higher risks and incidence of school absences, poor family cash flow, the experience of stress, and poor self-assessed health (Australian Institute of Health and Welfare, 2019). The historical policy of child removal motivated an emotional national apology in 2008 from the then Prime Minister Kevin Rudd, whose administration initiated the Closing the Gap strategy that sought to address aboriginal disadvantage.

First Nations people (Aboriginal and Torres Strait Islander) constituted 3.3 percent of the Australian population in 2016 (798,365 people). Contrary to popular perceptions, just 19 percent live in remote locations (AIHW, 2022). Historically, divided by vast physical distances into some 250 nations, languages, and cultures, First Nations people identify strongly with 'Country', the location of their birth, and their ancestors. The verbal tradition has carried origin explanations of land and people in 'songlines' enduring for millennia that create powerful identities and ties to place. It is estimated that there has been continuing aboriginal people and culture for some 65,000 years, the longest human culture. There is growing interest in traditional cultural practices of land stewardship, for example, cultural burning, which reduces fire load, at times calculated by complex balances of seasonal, climatic, and cultural rhythms. Despite an increasing valuation of First Nations cultures, acts of cultural vandalism remain common as exemplified by the destruction, of the 46,000-year-old Juurkan Gorge site by the mining giant Rio Tinto, in the Pilbarra Region of Western Australia (Hepburn, 2020). Regrettably, such acts are supported by Australian legislation and are common in furthering the development of the extractive industries in Australia and are clear examples of continuing legal and policy structures that perpetuate oppressive and paternalistic outcomes.

First Nations people experience complex patterns of multiple disadvantage and social exclusion. Income inequality is particularly marked. In 2018–2019, 40 percent of First Nations people were in the lowest 20 percent of the Australian income distribution. Just 8 percent were in the top 20 percent compared with 22 percent of the general population. The median income for the general population was 65 percent higher than that of the First Nations people. The highest median incomes (weekly) were found in the cities (AUD600), with those in 'very remote' areas having a median income of AUD350 (Australian Institute of Health and Welfare, 2021).

In the relational aspects of social exclusion, the multi-dimensionality of aboriginal poverty is also notable. Whilst there have been improvements over the last two decades in the rates of specific disadvantages, including life expectancy, child mortality, and educational attainment, the gaps between First Nations people and the wider Australian population remain significant. For example, for male life expectancy, the gap is currently 8.6 years compared with the non-indigenous population (Commonwealth of Australia, 2020a). This is despite an improvement of 4.1 years in the decade to 2018. Smoking and obesity rates are higher than for the general population, with cancer the primary cause of death, particularly respiratory and digestive system cancers. Two headline statistics also illustrate complex patterns of social exclusion. First, despite representing a small proportion of the overall Australian population, First Nations people disproportionately constitute 28 percent of the prison population. Despite a *Royal Commission into Aboriginal deaths in custody* (Johnson, 1991), this remains a significant problem for aboriginal communities and families, as incarceration causes cultural and family breakdown and none of the recommendations of the Royal Commission have been put in place.

A second notable experience of many aboriginal families is the removal of children into the care system. High levels of removal have promoted growing concerns of a further 'stolen generation' of aboriginal children. In 2016, the overall national rate of those in 'out of home care' settings was 5.5 per 1,000, this compares disproportionately with the rate of 56.6 per 1,000 among the First Nations population – almost ten times the national level (O'Donnell et al., 2019). Whilst there are factors – such as the deep poverty experienced by many aboriginal families, poor housing conditions, overcrowding, and alcohol abuse – that may partially explain these discrepancies, there are also other factors – including poor access to support services, cultural competence of agencies to understand the non-nuclear family structure of First Nations communities and, inevitably, elements of institutional racism in the services provided (Socha, 2020).

Housing conditions in remote First Nations communities are characterised by poor quality housing and high levels of overcrowding, a pattern most clearly revealed in the instances where COVID-19 has reached aboriginal communities, causing high rates of transmission. The impacts of overcrowding are particularly notable in rural and remote aboriginal communities (Lowell et al., 2018). With housing often designed for the European nuclear family, little attention has been paid historically to the cultural obligations to shelter kin. Housing agencies often allow overcrowding in recognition of cultural needs, for example during 'sorry time', an extended period of mourning following a death. The impact of general overcrowding on health, particularly of children is considerable, with childhood illnesses such as rheumatic fever and otitis media, long eradicated in modern health systems, continuing to cause lifelong health issues. Poor housing also impacts First Nations people in urban areas, where similar impacts on health are noted (Andersen et al., 2016).

On a positive note, there has been a notable improvement in the number of First Nations children completing Year 12 education. A completion rate of 45 percent in 2008 reached 66 percent in 2018–2019. Despite this considerable improvement, however, there remains a gap of 25 percent in comparison with the non-indigenous population. Furthermore, overall statistics obscure the considerable variation between urban and 'very remote areas', where the rate of year 12 completion remains at a very low 38 percent (AHRC, 2020). Consequently, educational disadvantage remains a core issue that impacts on future employability and quality of life. As with many issues impacting First Nations communities, analysis and policy have been framed from a 'deficit' perspective (Hogarth, 2017), with little recognition of the cultural and racialised practices discussed earlier. To date, there has been little policy recognition of the overwhelming evidence for the value of co-designing learning with local populations, which might help ensure that schooling connects pupils to culture and country, rather than assuming a cultural assimilation with the dominant European culture (Lowe et al., 2019; Louth et al., 2021). There is, however, some prospect of greater influence from co-design approaches with the 2020 Closing the Gap Agreement (Commonwealth of Australia, 2020b), which now includes a 'Coalition of Peaks' representing a wide range of First Nations people's interests.

In this brief review, we can see, that in the 'triangle of poverty' comprising health, education, and housing, the Australian First Nations people experience a wide range of disadvantages, despite some improvement since the instigation of the initial Closing the Gap programme in 2007–2008 only two of the seven targets were on track. We should, however, not accept the challenges faced by the First Nations community in Australia as a deficit model of their culture and economic capacity. These issues are the consequence of enforced cultural change, political domination, and economic exploitation by a colonising power. It is continued in contemporary practices of cultural colonisation that dominate relationships between First Nations and the dominant white European settler culture. The story of Australia's First Nations people is one of resilience and resistance. This has most recently been expressed in the Uluru Statement from the Heart, a proposal calling for a 'First Nations voice enshrined in the Constitution' (Uluru Statement, 2017). Instantly rejected by government, the Statement and associated call for a truth and reconciliation Makarrata commission remains a powerful expression of First Nations unity that cannot be ignored.

This brief review of the Australian context could be replicated in all settler dominated societies of the Anglophone world, where similar patterns of historical partial genocide, occupation, cultural repression, and political and institutional racism have been sustained for centuries, and which continue to fix indigenous populations in complex relationships between poverty and place. The cultural trauma of this experience will continue to damage the prospects of future generations of indigenous people, unless major reform and reparation, coupled with comprehensive political recognition, is achieved. Progress will be dependent on comprehensive constitutional recognition of First Nations people, treaty recognising rights, and a truth and reconciliation process in which the dominant culture faces its past.

This chapter has explored the wider context of national and international socio-economic relationships in which our case studies in Part Two are situated.

We can see that, what happens at the local level of place is conditioned and mediated by global social and political perspectives that dominate government and associated economic and welfare systems. The transition from a Keynesian economic order to a neoliberal paradigm had local consequences for the diverse places in our Deep Place studies. Understanding historical patterns of development and historical relationships at the national and global level can also inform our futures, and the actions that will be required to create 'places of hope' that address the environmental and social crisis that is the concern of this book.

References

Abel-Smith B and Townsend P (1965) *The Poor and the Poorest. A New Analysis of the Ministry of Labour's Family Expenditure Surveys 1953–54 and 1960*.LSE Occasional Papers in Social Administration. London: London School of Economics.

Adamson D (1998) 'The spatial organization of difference and exclusion in a working class community', *The Spokesman: Full Employment: A European Appeal*, 64: 141–152.

Adamson D (2001) 'Social segregation in a working-class community: economic and social change in the South Wales coalfield', pp. 101–128 in: G Van Guys, H De Witte and P Pasture (eds.) *Can Class Still Unite? The Differentiated Workforce, Class Solidarity and the Trade Unions*. Aldershot: Ashgate.

Adamson D (2008) 'Still living on the edge', *Contemporary Wales*, 21: 47–66.

Adamson D (2010) *The Impact of Devolution: Area-based Regeneration Policies in the UK*. York: JRF.

Alston P (2018) *Statement on Visit to the United Kingdom by Professor Philip Alston, United Nations Special Rapporteur on Extreme Poverty and Human Rights*. New York: United National Human Rights Office of the High Commissioner.

Andersen M J, Williamson A B, Fernando P, Redman S and Vincent F (2016) 'There's a housing crisis going on in Sydney for Aboriginal people, focus group accounts of housing and perceived associations with health', *BMC Public Health*, 16(1): 1–10. https://doi.org/10.1186/s12889-016-3049-2

Armstrong D V (2019) 'Capitalism and the shift to sugar and slavery in mid-seventeenth-century Barbados', *Historical Archaeology*, 53(3–4): 468–491. https://doi.org/10.1007/s41636-019-00213-8

Australian Institute of Health and Welfare (2019) *Children living in households with members of the Stolen Generations*. Cat No. IHW214. https://www.aihw.gov.au/getmedia/a364d8f1-eeee-43c3-b91e-0fb31ebecf30/AIHW214-Children-and-Stolen-Generation.pdf.aspx?inline=true

Australian Institute of Health and Welfare (2021) *Indigenous income and finance: A snapshot*. https://www.aihw.gov.au/reports/australias-welfare/indigenous-income-and-finance

Ball M and Maginn P J (2005) 'Urban change and conflict: evaluating the role of partnerships in urban regeneration in the UK', *Housing Studies*, 20(1): 9–28. https://doi.org/10.1080/0267303042000308705

Bangura Y (2019) 'Convergence Is Not Equality', *Development and Change*, 50(2): 394–409. https://doi.org/10.1111/dech.12489

Boettke P J and Smith D J (2016) 'Evolving views on monetary policy in the thought of Hayek, Friedman, and Buchanan', *Review of Austrian Economics*, 29(4): 351–370. https://doi.org/10.1007/s11138-015-0334-8

Bowman A, Erturk I, Froud J, Sukhdev J, Law J, Leaver A, Moran M and Williams K (2014) *The End of the Experiment: From Competition to the Foundational Economy*. Manchester: Manchester University Press.

Breman J, Harris K, Lee C K and Linden M van der (eds.) (2019) *The Social Question in the Twenty-First Century? A Global View*. Oakland: University of California Press.

Campbell B (1993) *Goliath: Britain's Dangerous Places*. London: Methuen.

Campbell S (2021) 'An appeal to supersede the slave trade triangle in English museums', *Atlantic Studies*. https://doi.org/10.1080/14788810.2021.1913969

Chancel L, Piketty T, Saez E and Zucman G (2022) *World inequality report*. World Inequality Lab. https://wir2022.wid.world/www-site/uploads/2021/12/WorldInequality Report2022_Full_Report.pdf

Cipolla C and Howlett-Hayes K (2015) *Rethinking Colonialism: Comparative Archaeological Approaches*. Gainesville: University Press of Florida.

Clasen J (2002) 'Unemployment and unemployment policy in the UK: increasing employability and redefining citizenship', pp. 59–74 in: J Goul Anderson, J Clasen, W van Oorschot and O Halvorsen (eds.) *Europe's New State of Welfare. Unemployment, Employment Policies and Citizenship*. Bristol: Policy Press.

Cole Julio H (2007) 'Milton Friedman, 1912–2006', *The Independent Review*, 12(1): 115–128.

Commission A H R (2020) *Closing the gap report 2020*. Australian Government. https://ctgreport.niaa.gov.au/sites/default/files/pdf/closing-the-gap-report-2020.pdf

Commission on Urban Priority Areas (1985) *Faith in the City: A Call for Action by Church and Nation. The Report of the Archbishop of Canterbury's Commission on Urban Priority Areas*. London: Church House Publishing.

Commonwealth of Australia (2020a) *Overcoming Indigenous Disadvantage. Key Indicators 2020*. Canberra: Productivity Commission, Steering Committee for the Review of Government Service Provision.

Commonwealth of Australia (2020b) *National agreement on closing the gap*. Commonwealth of Australia. https://coalitionofpeaks.org.au/wp-content/uploads/2021/04/ctg-national-agreement-apr-21-1-1.pdf

Cribb J, Waters T, Wernham T and Xiaowei X (2021) *Living standards, poverty and inequality in the UK: 2015–2016 to 2020–2021*. Institute of Fiscal Studies. https://ifs.org.uk/uploads/R194-Living-standards-poverty-and-inequality-in-the-UK-2021.pdf

Crisp R, Pearson S and Gore T (2015) 'Rethinking the impact of regeneration on poverty: a (partial) defence of a "failed" policy', *Journal of Poverty and Social Justice*, 23(3): 167–187. https://doi.org/10.1332/175982715X14443317211905

Crossley S (2017) 'The "official" social justice: an examination of the Coalition government's concept of social justice', *Journal of Poverty and Social Justice*, 25(1): 21–33. https://doi.org/10.1332/175982717X14842282011532

Deeming C (2015) 'Foundations of the workfare state – reflections on the political transformation of the welfare state in Britain'. *Social Policy and Administration*, 49(7): 862–886. https://doi.org/10.1111/spol.12096

Dickerson A and Popli G (2018) 'The Many Dimensions of Child Poverty: Evidence from the UK Millennium Cohort Study', *Fiscal Studies*, 39(2): 265–298. https://doi.org/10.1111/1475-5890.12162

Donington K, Hanley R and Moody J (2022) 'Introduction', pp. 1–18 in: K Donington, R Hanley and J Moody (eds.) *Britain's History and Memory of Transatlantic Slavery: Local Nuances of a "National Sin"*. Liverpool: Liverpool University Press.

Douglas H and Walsh T (2013) 'Continuing the stolen generations: child protection interventions and indigenous people', *International Journal of Children's Rights*, 21(1): 59–87. https://doi.org/10.1163/157181812X639288

Fitzpatrick S and Pawson H (2016) 'Fifty years since Cathy Come Home: critical reflections on the UK homelessness safety net'. *International Journal of Housing Policy*, 16(4): 543–555. https://doi.org/10.1080/14616718.2016.1230962

Freeman D (2020) *Can Globalization Succeed? A primer for the 21st century*. London: Thames and Hudson.

Fukuyama F (1992) *The End of History and the Last Man*. New York: Free Press.

Gazeley I, Gutierrez Rufrancos H, Newell A, Reynolds K and Searle R (2017) 'The poor and the poorest, 50 years on: evidence from British Household Expenditure Surveys of the 1950s and 1960s', *Journal of the Royal Statistical Society. Series A: Statistics in Society*, 180(2): 455–474. https://doi.org/10.1111/rssa.12202

Giddens A (1998) *The Third Way. The Renewal of Social Democracy*. Cambridge: Polity Press.

Glasmeier A, Martin R, Tyler P and Dorling D (2008) 'Editorial: poverty and place in the UK and the USA', *Cambridge Journal of Regions, Economy and Society*, 1: 1–16. https://doi.org/10.1093/cjres/rsn004

Goldthorpe J, Lockwood D, Bechhofer F and Platt J (1969) *The Affluent Worker in the Class Structure*. Cambridge: Cambridge University Press.

Gray S (2007) 'The elephant in the drawing room: slavery and the "stolen wages" debate', *Australian Indigenous Law Review*, II(3): 30–53.

Hall C (2020) 'The slavery business and the making of "Race" in Britain and the Caribbean', *Current Anthropology*, 61(S22): S172–S182. https://doi.org/10.1086/709845

Hepburn S (2020) 'Rio Tinto just blasted away an ancient Aboriginal site. Here's how that was allowed', *The Conversation*. https://theconversation.com/rio-tinto-just-blasted-away-an-ancient-aboriginal-site-heres-why-that-was-allowed-139466

Hickel J (2017) 'Is global inequality getting better or worse? A critique of the World Bank's convergence narrative', *Third World Quarterly*, 38(10): 2208–2222. https://doi.org/10.1080/01436597.2017.1333414

Hills J (2011) 'The changing architecture of the UK welfare state', *Oxford Review of Economic Policy*, 27(4): 589–607. https://doi.org/10.1093/oxrep/grr032

HM Government (2021) *Levelling up Fund: Prospectus* (Issue March). Crown Publisher. London

Hogarth M (2017) 'Speaking back to the deficit discourses: a theoretical and methodological approach', *Australian Educational Researcher*, 44(1): 21–34. https://doi.org/10.1007/s13384-017-0228-9

Inikari J and Engerman S (1992) 'Gainers and losers in the Atlantic Slave Trade', pp. 1–21 in: J E Inikari and S L Engerman (eds.) *The Atlantic Slave Trade: Effects on Economies, Societies and Peoples in Africa, The Americas and Europe*. Durham, NC and London: Duke University Press.

Johnson E C (1991) *Royal Commission into aboriginal deaths in custody*. http://www.austlii.edu.au/au/other/IndigLRes/rciadic/national/vol1/1.html

Jordan B (2008) 'The place of "place" in theories of poverty: mobility, social capital and well-being', *Cambridge Journal of Regions, Economy and Society*, 1: 115–129. https://doi.org/10.1093/cjres/rsm002

Kelton S (2020) *The Deficit Myth: Modern Monetary Theory and How to Build a Better Economy*. London: John Murray.

Lash S and Urry J (1987) *The End of Organised Capitalism*. Bristol: Policy Press.

Levitas R (1998) *The Inclusive Society? Social Exclusion and New Labour*. Basingstoke: MacMillan.

Louth S, Wheeler K and Bonner J (2021) 'Long-lasting educational engagement of Aboriginal and Torres Strait Islander people: where are the Ghundus (children)? A longitudinal study', *Australian Journal of Indigenous Education*, 50(1): 107–115. https://doi.org/10.1017/jie.2019.14

Lowe K, Tennent C, Guenther J, Harrison N, Burgess C, Moodie N and Vass G (2019) 'Aboriginal voices: an overview of the methodology applied in the systematic review of recent research across ten key areas of Australian Indigenous education', *Australian Educational Researcher*, 46(2): 213–229. https://doi.org/10.1007/s13384-019-00307-5

Lowell A, Maypilama L, Fasoli L, Guyula Y, Guyula A, YunupiLatin Small Letter Engu M, Godwin-Thompson J, Gundjarranbuy R, Armstrong E, Garrutju J and McEldowney R (2018) 'The "invisible homeless" – challenges faced by families bringing up their children in a remote Australian Aboriginal community', *BMC Public Health*, 18(1): 1–14. https://doi.org/10.1186/s12889-018-6286-8

Lyndon S (2019) 'Troubling discourses of poverty in early childhood in the UK', *Children and Society*, 33(6): 602–609. https://doi.org/10.1111/chso.12354

MacInnes T, Aldridge H, Bushe S, Kenway P and Tinson A (2013) *Monitoring poverty and social exclusion 2013*. https://doi.org/10.1111/j.1365-2214.2004.406_5.x

MacLeavy J (2011) 'A "new politics" of austerity, workfare and gender? the UK coalition government's welfare reform proposals', *Cambridge Journal of Regions, Economy and Society*, 4(3): 355–367. https://doi.org/10.1093/cjres/rsr023

McBride J, Smith A and Mbala M (2018) '"You End Up with Nothing": The Experience of Being a Statistic of "In-Work Poverty" in the UK', *Work, Employment and Society*, 32(1): 210–218. https://doi.org/10.1177/0950017017728614

Milanovic B (2016) *Global Inequality: A New Approach for the Age of Globalization*. Cambridge, MA: Belknap Press.

Möller C (2021) 'Discipline and feed: food banks, pastoral power, and the medicalisation of poverty in the UK', *Sociological Research Online*, 26(4): 853–870. https://doi.org/10.1177/1360780420982625

Monbiot G (2016) *How Did We Get into This Mess: Politics, Equality, Nature*. London and Brooklyn: Verso.

Murray C (1990) *The Emerging British Underclass*. London: Institute of Economic Affairs.

Oatley N (2000) 'New labour's approach to age-old problems: renewing and revitalising poor neighbourhoods – the national strategy for neighbourhood renewal', *Local Economy*, 15(2): 86–97. https://doi.org/10.1080/02690940050122659

O'Donnell M, Taplin S, Marriott R, Lima F and Stanley F J (2019) 'Infant removals: the need to address the over-representation of aboriginal infants and community concerns of another "stolen generation"', *Child Abuse and Neglect*, 90(February): 88–98. https://doi.org/10.1016/j.chiabu.2019.01.017

Philo C (1995) *Off the Map: The Social Geography of Poverty in the UK*. London: Child Poverty Action Group.

Piketty T (2013) *Capital in the Twenty-First Century*. Cambridge, MA and London: Belknap Press.

Piketty T (2021) *Time for Socialism: Dispatches from a World on Fire 2016–2021*. New Haven, CT and London: Yale University Press.

Rollings N (2013) 'Cracks in the post-war Keynesian settlement? The role of organised business in Britain in the rise of neoliberalism before Margaret Thatcher', *Twentieth Century British History*, 24(4): 637–659. https://doi.org/10.1093/tcbh/hwt005

Room G (1995) 'Poverty in Europe: Competing Paradigms and Analysis', *Policy and Politics*, 23(2): 103–113.

Said E (1978) *Orientalism*. New York: Pantheon Books.

Savage M, Devine F, Cunningham N, Taylor M, Li Y, Hjellbrekke J, LeRoux B, Friedman S and Miles A (2013) 'A new model of social class? Findings from the BBC's Great British class survey experiment', *Sociology*, 47(2): 219–250.

Savage M (2014) 'Piketty's challenge for sociology', *British Journal of Sociology*, 65(4): 591–606. https://doi.org/10.1111/1468-4446.12106

Scarman L (1986) *The Scarman Report: The Brixton Disorder 10th-12th April 1981*. Harmondsworth: Penguin.

Shaw K and Robinson F (2010) 'UK Urban regeneration policies in the early twenty-first century: Continuity or change?', *Town Planning Review*, 81(2): 123–150. https://doi.org/10.3828/tpr.2009.31

Socha A (2020) 'Addressing institutional racism against aboriginal and Torres strait islanders of Australia in mainstream health services. Insights from aboriginal community controlled health services.' *International Journal of Indigenous Health*, 15(1);: 291–303.

Standing G (2011) *The Precariat: The New Dangerous Class*. London and New York: Bloomsbury Academic.

Taylor-Robinson D, Rougeaux E, Harrison D, Whitehead M, Barr B and Pearce A (2013) 'The rise of food poverty in the UK', *BMJ* (Online), 347(December): 2–3. https://doi.org/10.1136/bmj.f7157

Uluru Statement from the Heart (2017) Accessed 28 January 2022: https://ulurustatement.org/the-statement/view-the-statement/

United Nations Development Programme (UNDP) and Oxford Poverty and Human Development Initiative (OPHDI) (2020) Global Multidimensional Poverty Index 2020. Charting pathways Out of Multidimensional Poverty. Achieving the SDGs. Accessed 17 December 2021: http://hdr.undp.org/sites/default/files/2020_mpi_report_en.pdf

Wincott D (2013) 'The (Golden) age of the welfare state: interrogating a conventional wisdom', *Public Administration*, 91(4): 806–822. https://doi.org/10.1111/j.1467-9299.2012.02067.x

4

ENVIRONMENT AND PLACE

Understanding the socio-natural relations of the Anthropocene

This chapter considers the active process of marginalisation and peripheralisation in the context of unequal social and natural rights. It seeks to address how we may develop a critically normative approach to nature and place given the Anthropocene. There is a need to move beyond the treatment of the environment as a compartmentalised and reductionist non-human space, or as a surface or set of layers to be protected or restored. Nature and environment are inherently common pool resources, but we are confronting a long process of metabolic rift, that is the regressive detachment of populations from their natures. In addition, the combination of carbonism and capitalism, especially in its most extreme and modern forms, continues to assymetrically engrain systems of natural property and bio-sphere rights into exclusive havens with strong boundaries and barriers to entry and participation. In fact, natural relations for the human population have increasingly become a positional good, dependent on status and position in society.

This creates a *double-environmental jeopardy problem* during the Anthropocene. On the one hand, as we shall outline below, carbonised capitalism tends to create boundaries around natures that tend to exclude large proportions of urban and indigenous people, under conditions of rapid recent historical (and privatised) urbanisation and population growth. On the other hand, whilst being excluded from either renewable or non-renewable exploitation and capture of those natures, these dispossessed populations are exposed directly to their very negative impacts in the form of a variety of pollution cocktails and the continuous re-production of social and economic vulnerabilities. This has, over the past 200 years or so, made the notion of metabolic rift increasingly toxic. The marginalisation of peoples' natures exacerbates their overall social and economic marginalisation; these two dynamics go hand-in-hand. They may be offset by the positional benefits

DOI: 10.4324/9781003221555-5

of globalisation for a time as we can create spatial fixes. The current approach to 'offsetting' our spatial fixes, however, is coming to an end as there are few places left to exploit or dump stuff without a rise in moral, social, and territorial consciousness. Hence, the need for the rise in not only inter-generational equity but territorial and environmental justice and freedoms.

Current mindsets in policy, governance, and academia are not resolving, but indeed re-enforcing the double jeopardy rock and a hard place problematic. They are realising the (rock) nature/environment nexus, and they are (partly) recognising the hard place narrative of the continued process of acceptable and legitimate social and economic marginalisation. We are, however, still at the stage of recognition not action on reversing and upending these processes. This chapter elaborates a new framework, which confronts these issues head on. The main conclusion of this chapter suggests that we need to reverse this false and artificially created inverse relationship into a positive relationship between natural engagements, practices, and rights. In this context, we argue that a focus on place becomes a key fulcrum for this reversal.

Key analytical issues covered in this chapter will include:

- The contestations of indigenous peoples and communities as they try to protect their customary rights against the forces of enclosure and privatised appropriation (see Chapter 3 for an account of these processes for the Australian First Nation community).
- The process of privatised mass (con)urbanisation.
- Recent examples of the regressive rentier privatisation of domestic and public space.
- A discussion on engaged and empowering social practices, and how they are created and sustained in and by communities and places.
- A discussion of how new and innovative community and place-based forms of practice and action can be extended.
- A consideration of how a transformational environmental agenda can empower people to take back their natures, and whether this implies a more distributed/ hybrid eco-capitalism that seeks to redistribute and re-orientate property and bio-sphere rights (see Chapter 10 for our discussion of 'regenerative collectivism').

Social nature as re-domestication or emancipation?

Current environmental policies and movements, whilst attempting to address and indeed re-dress the negative ecological effects of the last 200 years of industrialisation, tend to assume that the overriding social and economic structures that prevail can adapt without too much disturbance. As such, there is an in-built assumption in much environmental policy and thinking that environmental restoration and protection can be rolled out as a discrete sectoral policy whilst leaving much of the rest of societal and spatial organisation largely intact. This engrained and prevailing assumption is one of the main reasons why critical environmental thinkers calling for more integrated and systems thinking, have, largely, been unsuccessful in persuading policy-makers of the need for more fundamental change. *The current environmental crisis is thus a crisis of the prevailing and limited theories and practices of social and spatial development.* This failure has at least three important consequences:

1   It compartmentalises the environment into a series of black boxes or silos and allows rampant and increasingly unstable capitalism to continue business as usual.
2   The policies that emerge tend to constrain, domesticate and restrict public, territorial, and civil freedoms as well as re-enforce territorial and environmental injustice and inequalities. Urban intensification and privatisation of urban living space take place at the same time as human vulnerabilities of high densities become reproduced.
3   Growing social inequalities are manifested in increasingly asymmetric socio-natural relations. By adopting a weak form of ecological modernisation, emerging environmental policies over the past 50 years have tended to assume the further concentration of wealth and property rights, which now extend into not only financial mass and flows (see Harvey, 2021), but also are now extended into every mode of life involving the body, the bio-sphere and the ground and soils beneath us.

These consequences are leading to forms of social and cultural re-domestication of a hyper con-urbanised population whose environmental rights are increasingly squeezed. By continuing to separate the environment from its place-based nature and its in-built societal functioning, this further exacerbates the overall Anthropocentric conundrum: recognising that humans have caused the chaos, whilst assuming this can be resolved by technicist means that are largely divorced from everyday livelihoods. In this chapter, we explore these consequences using and applying a series of social-nature and place-based concepts that focus on the *nature-societal power relations of place.* These include domestication and enslavement; milieux, ambiance, and situationism; exposure and vulnerability; distanciation and struggles over empowerment; and enclosure and commoning. These concepts begin to help us understand socio-natural power relations in an increasingly (con)-urbanised world, where nature increasingly becomes distanciated from everyday life, or at best enclosed in ways that domesticate populations and reduce their empowerments over nature.

Global agglomeration as spatial development: a process of metropolitan acculturation and monopolies over nature

It is common to assume that since the second demographic transition, argued to have begun around 1950, the world has become less rural. By 2007, more than 50 percent of the world's population was urban, even though the developing world remains more rural than urban. The majority of countries, such as the USA, Canada, Sweden, Australia, China, and Japan, have experienced or are experiencing a severe rural population decline, with attractive urban living standards being seen as the major cause. This was especially the trend in newly industrialising and urbanising countries in East and South Asia, Latin America, and parts of Africa. The rural population of China has decreased by 45.9 percent during the past 35 years, whilst the figure in Brazil has almost reached 60 percent (Li et al., 2019).

Increasingly the burgeoning numbers living in urban areas, and many of those in the intermediate suburban areas surrounding expanding city regions, are experiencing a new process of *cosmopolitan domestication*. This has been a feature since early industrialisation in 19th century Europe, but by the middle of the 20th century, it became a dominant and mass demographic shift. By the start of the 21st century, the scale of con-urbanisation was unprecedented, and increasingly dense urban populations have become more dependent upon distanced rural resources (such as food, energy, and minerals). Consequently, people are continuing to lose their rural roots, identities, and ways of living as they become re-domesticated in urban post-natural environments, which make them increasingly dependent upon complex supply chains (see: Scott, 2017). Urban domestication tends to be seen as a sign of progress and progressive modernisation, symptomatic of the knowledge economy, and essential for economic growth. Urban dwellers have, however, lost the skills and abilities to self-procure and are increasingly reliant on purchasing commodified and manipulated goods and services.

The massive and historically unique process currently underway places rural development in a significantly re-located and critical position. The highly conglomerated and interconnected cities of China, for example, have become increasingly dependent upon rural-based resources, not least rocks and minerals. A hundred miles west of Shanghai the devastation of rural hills and mountains, quarried for their urban building materials, is immediately apparent. Around many more established cities (such as London or Sao Paulo), vastly extended food supply chains spread not only to their geographically close rural hinterlands, but also to internationalised importing platforms for fresh fruits and vegetables (such as the Sao Francisco Valley in North East Brazil, or the rural heartlands of Ghana). Rural people tend to move temporarily or permanently to these cities, but even when they remain in their rural communities many become closely affected by the urban and cosmopolitan demands for rural-based resources. We thus recognise that the age of urban cosmopolitanism comes with new human and cultural processes of rural development, urban re-domestication and, indeed rural enclosure. These processes have become a major feature of global rural development, but as Li et al. (2019, p. 140) argue:

> ...rural decline is not pre-destined. It is by the interactions between rural areas and the external environment that rural communities either grow,

decline or even vanish....rural communities of different geographical conditions, natural resource endowments and social relationships, as well as
people's values, attitudes and institutions will make different responses,
which finally lead to different evolution patterns and outcomes.

In the context of the ongoing process of urban domestication, significant differences are occurring between the expanding and densely living urban, and the
more sparsely populated rural. These differences include:

- Mortality in the South tends to be higher in rural areas than urban areas,
  largely resulting from higher mortality of children under the age of five.
  This stems from many rural areas having lower incomes, poorer nutrition,
  less clean water and sanitation, and fewer medical services than their urban
  counterparts (this is a mean observation, however, and it should be noted that
  urban domestication is recreating many health disadvantages) (ODI, 2016).
- Fertility in rural areas tends to be higher than urban areas. This may result
  from lower incomes, but may also be associated with agriculture (it is easier
  to combine child raising with farm work than with factory work).
- The combination of higher mortality and fertility can appear to delay the
  process of urban domestication, so that rates of natural population growth
  can be higher in rural areas; although most rural areas in the South experience out migration into the cities, such that these grow proportionately faster.
- Without migration the higher natural rates of population growth in rural
  areas would tend to ruralise rather than urbanise countries. Migration is
  therefore essential for continuing urban domestication and, consequently,
  current models of economic growth.
- Urban domestication is leading to higher levels of aggregate carbonised energy use. As urban domestication implies far higher levels of consumption,
  and correspondingly lower levels of direct participation in production (e.g. of
  energy and foods), surrounding rural hinterlands become critical for supply.
- As more people are locked into path-dependent lifestyles and consumption
  patterns, not of their own making, questions are raised about what effects
  this has: (i) on existing rural areas and peoples; (ii) on urban people's resilience or vulnerability; and (iii) on creating opportunities or obstacles for
  post-carbonised forms of transformation. In short, how will the current demographic fix play out once it is realised that the agglomerative economic
  and cultural advantages of the city no longer apply?
- Experiences of social inequality, poverty, and social exclusion are now rife
  in both urban and rural settings, but it remains a stark fact that migrations
  to the city are often an expression of desperation and distress in rural living
  standards, whilst return migration (especially in the North) is often associated with a quest to retain non-urban or suburban domesticated ways of
  living for the relatively rich and urban successful seeking to 'escape to the
  country' (Thompson, 1982).

Rural-urban relations are thus at something of a turning point: will the urbanisation trend that has risen from 20 percent in 1960 to 46 percent in 2014, continue in the face of wholesale transformations in our use of energy, material resources, and food supplies? What will be the main pathways towards more sustainable forms of rural and urban development? Will we undergo a third demographic transition with huge dispersals of now urban-domesticated populations re-occupying rural hinterlands? Will rural areas face new forms of urban and suburban-based enclosure, where their resources are appropriated under conditions of more extreme resource depletion for the purposes of satisfying stronger state and corporate hegemons?

Urban domestication, rural enclosure, and pathways to empowerment

We can thus begin to draw what some might regard as a new societal and cultural dichotomy between the urban and the rural, in the context of our current global and globalising ecological and socio-economic crisis. This does not suggest that the rural and the urban are disconnected. Quite the opposite in that they each increasingly feed off each other. The purpose here is to conceptually re-ask important sociological and geographical questions, such as: what is it like to be urban and what is it like to experience rurality? What is the nature of rurality in post-modern society when hyper-mobilities collide with new ecological vulnerabilities and social inequalities? Do these conditions engender a new transition towards, as Latour (2018)) suggested, re-territorialised processes as a reaction to the global and globalised ecological and economic crisis? Ironically, rural places again become places to put down roots, almost as an antidote to globalisation and ecological crisis.

It is possible to posit that the process of urban or metropolitan domestication has and continues to be a global and universal process affecting both urban and rural populations. Access to diverse consumption patterns, high levels of mobility, and the opportunities choice brings tends to be accounted as a global universal. In rural areas, however, the dangers are that the very demands that such (urban) domestications place on natural resources significantly affect rural areas. Clearly, this is most commonly witnessed by the recent round of rural land grabbing – or the *new enclosures*. Since 2007, an estimated 220 million hectares has been acquired by foreign investors in the Global south. Intensive forms of increasingly vulnerable urban domestication are being offset by more intensive and profound forms of financialised enclosure in many rural regions, especially those areas (such as in Africa and Latin America) where pre-existing rules and governance regimes are relatively weak to resist these financialised pressures.

Whilst much has been made of the recent rounds of land-grabbing as new expressions of hyper-financialisaton and the expression of accumulation by dispossession (Harvey, 2007) by extracting and exporting primary natural resources, it is also the case that this is a by-product of the intensive process of urban domestication being experienced in countries like China, South Asian countries and (food and water insecure but rapidly urbanising) Middle Eastern Oil States. Expanding and densely populated cities rely upon vast and expanding ecological and indeed rural footprints (it is argued that London's is the equivalent size of Spain). Cities lie at the epicentres of vast and complex supply chains for natural and processed resources. *Intensive forms of urban domestication are going hand-in-hand with selective and intensive forms of new rounds of rural enclosure, now at a globalised scale.* Just as the early urban manufacturing cities demanded former rural labour and resource commodification and enclosures in the 18th and 19th centuries, in the 21st this has massively intensified with networks of mega-cities demanding globally extractive enclosures across the globe (see for instance the development of China's 'belt road' initiative). The growth of mega-urbanisation is coming at an increasingly high material and ecological cost.

If we conceive of mega-urban domestication and rural enclosures as two sides of the same coin at least in hyper-capitalist and carbonist hegemonic systems, then it also begins to raise questions of who is enslaving whom, and who is empowering whom? Or, to put it another way, what types of power and empowering relations are these interdependencies creating? Not all rural areas are caught up in webs of the extractive enclosure, with many (in Western Europe, for instance), it would seem able to create more autonomous systems of empowerment that enable endogenous social and economic development to take hold. Similarly, amid the high-rise city, we find counter-hegemonic forces developing, not least around access to sustainable foods, housing, and energy schemes. Indeed, the spectre of post-carbonism shows some signs of encouraging post-capitalist, and certainly post-neoliberal modes of social organisation and profit sharing that are counter-hegemonic – they create the social conditions for social empowerment over consumerist 'enslavement'.

We can conceive of this as not so much as accumulation by dispossession, but as new forms of empowerment by association. The latter tendency is becoming a more dynamic force in many urban and rural settings and is creating innovative conditions for social action and sustainability strategies in many rural regions. Empowerment by association can be considered a transformational force, not least around agro-ecology and new empowered urban food consumption. People can, as the transition from capitalist carbonism accelerates, find the space and social power to create new associations and practices based upon different and more emancipated forms of organisation (also considered in Chapters 2 and 10). Hence the city is not just a site of domestication and disempowerment, just like the countryside is not just a place of enslavement or enclosure. Both the complex unfolding of the tendencies for domestication and enclosure thus need to be modified and qualified by considering how new sites and networks of association and empowerment can take hold. There are many contradictions and dilemmas for contemporary rural peoples and areas. We wish, in a non-reductionist and open-ended way, to explore these tendencies as the contradictions and dilemmas play out.

Reflecting on urban places to understand rural places: restricting social ambience and milieu

In the contemporary (especially con-urban) world the boundaries between freedom and captivity are far more complex than recognised in antique periods. In the chapter that follows we discuss the implications of this within the context of agri-food systems and metabolism, but the questions of enslavement and empowerment through complex digital and physical forms of control are no less relevant. Take, for example, the relatively new use of digital facial recognition technologies. Whilst in the West these are increasingly controlled by security agencies and a small group of digital and social media firms, in the East, they are becoming literally embodied in civil society actions by communist states (see Lanchester, 2019; and Strittmatter, 2019). The progress in facial recognition technologies and big data are being used not just for surveillance of popular unrest, but are also central to the pilot trials of what are termed social credit systems. This is far extending the west's use of financial credit ratings into what is seen by the state as socially desirable behaviour.

The rise of urban-based non-communicable diseases, diet-related human health risks (discussed in Chapter 5; also see Lee et al., 2020) and more intensive expressions of ambient air and water pollution are relatively new forms of urban enslavement. These are built upon density and convenience-driven urban re-domestication of household and work patterns. It is, for example, argued that pollution of water systems in urban neighbourhoods creates conditions of anti-microbial resistance and rising infertility. Furthermore, high density and high-rise living bring a wide range of additional health risks, particularly as income and social inequalities increase (see Bai et al., 2012, discussed in Chapter 3).

Air pollution, for example, has now re-emerged as a major global ambient urban killer – in the USA it is estimated that 200,000 people a year die as a result of such pollution. Combinations of outdoor and indoor air pollution caused one in every nine deaths globally in 2016, far more than those caused by malaria, malnutrition, or alcohol (Gardiner, 2019). In February 2020, as COVID-19 killed 2,714 people that month, 800,000 died from the effects of either outside or inside air pollution (see Gardiner, 2019). Overall, 10 million die from the effects of air pollution each year, and about 20,000 on an average day, four times the official worldwide COVID-19 death toll in 2020. There is, for instance, a close spatial correlation between children living in poverty and dirty air in the UK. The five London boroughs ranking highest for child poverty also rank highest for air pollution. In the 11 UK (urban) local authorities that currently exceed the WHO guideline limit for fine particle pollution (PM2.5 of an annual mean of ten micrograms per cubic metre of air), an average of 39.5 percent of children live in poverty – significantly higher than the national average of 31 percent. Overall, about 6.7 million children are living in areas of the UK where air pollution has breached legal limits, of which 2 million are also living in poverty (see Gardiner, 2020). This is not just a case of deepening social and environmental inequality, as we shall see below, it is a process of ecological enslavement and reductions in natural freedoms.

In the context of climate change urban living is, for many, an enslavement process involving high levels of metabolic and biological vulnerability, created and re-produced by an increasingly dense and polluting ambience, which restricts dwellers' environmental rights to natural and (what should be) common-pool resources (see further discussion in Chapter 5). This is exacerbated by the growing privatisation of green spaces in cities, and the largely unregulated nature of air and water pollution. *This is a form of ecological and biological urban enslavement.* The growing and increasingly recognised pathologies of urban life are also leading to a variety of rebound and interactive effects in rural development, and we therefore need to better understand the new dynamic linkages between the urban and the rural at a time of ecological and economic crisis.

Land and property rights: the dialectics of enclosure and commons

Issues of land rights, control, and use of the land and the biosphere are growing in significance as a result of the need to reduce and remove carbonised systems of production and supply. These issues are also relevant to the apparent requirement of established economic interests to secure land rights, in order to maintain profits in the context of unsustainable resource consumption practices. The control of rural land has thus become a fulcrum in the intense competition between de-peasantisation and re-peasantisation, as the contestations between traditional and endogenous landed communities face new rounds of the pressure of dispossession. This is clearly taking place in what have formerly been seen as global commons (like the Arctic and Greenland ice cap) or the indigenous lands of Northern Canada, Australia, or Brazil, which are under pressure for resource exploitation for forest products, precious minerals, and tar sands. In Latin America and Africa, global agri-business and mineral firms attempt to dispossess indigenous communities. Even in Europe, the privatisation of former public land is rapidly increasing.

These tensions are by no means new phenomena, as they raise the spectre of a re-formed and re-newed agrarian question, that is: *how will, or can local communities and landholders resist and build more sustainable pathways of rural development amidst such exogenous and appropriating forces?* This new land question is at the centre of debates concerning severe global resource depletion and the urgent search for post-carbonised alternatives, as well as issues relating to the rights of existing land (and biospheric resources) holders and users. In an increasingly urbanised and cosmopolitan world, it must also consider what rights non-land (urban) holders have to access the natural resources they require for survival. We can no longer frame the agrarian question, as earlier thinkers did, around the protection or projection of narrowly defined owner and occupier rights. The new land question is part of the global problem of how to allocate and govern natural resource rights.

These issues will be explored with reference to contemporary experiences in Europe and Latin America below, but it is necessary to first explore the historical antecedents of the new agrarian question more fully and to position this analysis within a more financialised contemporary context.

The new agrarian question: the new political ecology of resource governance

Since the 2007–2008 financial crisis, we have entered a constant dialectical process of economic, political, and ecological crisis. This has led to a radical shake up of political and sociological structural moorings, which is threatening established democratic and institutional structures, as well as longstanding regulatory processes developed to protect civil societies from economic and ecological vulnerability. These convergences are of anthropogenic significance, and there is now an urgent need to identify ways of addressing these threats (see Purdy, 2015; Marsden, 2018). Whilst other theorists have discussed the post-political tendencies in recent political ecology, we emphasise the need to create a *politicalised* political ecology. In this context, we need to address the issue of new agrarian transformations: from neoliberalist late-capitalist agrarianism to a potential post-neoliberalist and post-capitalist set of diverse conditions.

Rural societies, and their increasingly diversified and distributed agri-food/energy forms, are at the forefront of this highly contested and dialectical process (see Marsden and Ruscinska, 2019). This is of such profundity as to parallel what Kautsky called Die Agrarfrage (The Agrarian Question) at the start of the 20th century. Kautsky's problematic project infuriated the more conventional Leninist oligarchy of agrarian Marxism of the time, by asking: 'whether and how capital is seizing hold of agriculture, revolutionising it, making old forms of production and property untenable and creating the necessity for new ones?' (1899, p. 12). In the midst of rapid, if uneven, capitalist and carbonised industrialisation and urbanisation, and the new imperialist globalisation of agriculture, Kautsky's main concern was the continued role of pre-capitalist and non-capitalist forms of agriculture in capitalist societies (Friedmann and McMichael, 1989). The limiting and spatially variable nature of the soil did not, of course, stop concentrated capital (in the west) or the authoritarian state (in the East) from attempting to penetrate and transform distributed peasant production. This has been a continuous process of contested transformation throughout the 20th century (see Goodman et al., 1987).

Kautsky's legacy has new credence today as he gave serious consideration to the combination and distinctiveness of the material and biological worlds, and, specifically, the immovability of land and risks associated with biologically based agrarian domestication and climate vagaries over time and space. This unleashed over the succeeding century a continuing debate about explaining the persistence and empowerment of what seemed to be the anomalous structural location of the peasantry (see Marsden, 2017). So why are these interpretations of value today given our current and combinative sets of crisis rural development conditions, and why is it valuable to critically re-insert them into our interpretations of crisis over a century later? Firstly, we believe there is a need for a new neo-Kautskian viewpoint within a radically different set of external conditions from those that Kautsky experienced. Secondly, we can suggest some theoretical continuities, not least with his abiding theme, the *longue duree* distinctiveness of natural-social material production processes and practices in the face of advanced (corporate) capitalist penetration.

Changing conditions: theoretical continuities and additions

Agri-food systems are in ecological and economic crisis. Conventional agrarian capitalism shows all the signs of running its course throughout much of the 20th century, and it has been unable to overcome the distinctive organic and metabolic nature of production. Indeed, it has resorted to high and increasingly abstract forms of financialisation that are, in the long term, doomed to failure. Moreover, whilst agrarian capitalism predominates in many parts of the world (see Greenberg's [2018] discussion of appropriated intensive production in California's San Joaquin Valley) and continues to privatise rights and access to formerly common resources in much of Africa and Latin America, overwhelming evidence demonstrates that this has not lead to positive eco-system services or well-being outcomes (Rasmussen, 2018). We can, therefore, argue that in terms of its endogenous features (super exploitative labour, social and animal welfare, destruction, and extinction of biodiversity), and a range of wider external (landscape) factors (highly volatile markets, financialisation, consumer boycotts, and drought), the industrialised model of agri-food is confronting a far more serious array of Kautskian constraining factors today. We may also argue that the ecological world has closed in for conventional agrarian capitalism, as formerly cheap resources (labour and raw material like phosphates, minerals, etc.) have become harder to procure, and their negative externalities harder to spatially fix and legitimise. In this sense, the 21st-century capitalist world ecology is, after a series of successive Imperial, Fordist and corporate, neoliberalised and financialised globalisations (see Jason Moore, (2015) a much smaller and more vulnerable place.

Human-induced climate change and the negative effects of intensive agricultural capitalism did exist and were identified prior to the end of the 19th century (see Kropotkin, 1904; Davis, 2018), but this was largely ignored or denied until the final decades of the 20th century. This denial was based upon the assumption that agricultural intensification could continue unabated in world development, not least because technological advances would continue to deliver gains in productivity and continue to overcome any natural obstacles (at least for the foreseeable future). Meanwhile, wider and deeper neoliberalised globalisation processes (especially from the 1980s onwards) were continuing to allow capital and the state to create distanced spatial fixes in food production and consumption. These conveniently located the ecological externalities far away from Northern consumers who were increasingly environmentally and health conscious about their food intake and provisioning (Morgan et al., 2006). Towards the end of the 20th century and the beginning of the 21st century, these conditions continued to provide the means by which Kautsky's concerns and ecological crisis and climate change more broadly, could be conveniently denied and side-lined.

In many places, the political conditions have shifted along with the decline of state socialism, the rise of right-wing populism, and new trade wars. Such forms of disruptive governance are also affecting the demise of agrarian capitalism. Greenberg (2018/2019), for instance, documents not only the ecological crisis in the San Joaquin, but also the neo-right deportations of Mexican labour and the related humanitarian crisis leading to structural shifts in agri-food. Greenberg argues:

> As things stand, there is a labour shortage the magnitude of which hasn't been seen in the last ninety years. It has prompted growers to rip out labour intensive fruits like table grapes and plant almond trees, which require few workers. Housing costs, especially in the Coastal Valley, have made it even harder to attract and keep workers. In recent years millions of dollars of unpicked crops have been ploughed under or left to rot in the fields.
>
> *(p. 93)*

Whilst there did appear to be a long period of plenty regarding food supplies and choices, at least in the advanced countries of the North, since 2007–2008 there has been a severe rise in food insecurities and associated global political turmoil. This is clearly demonstrating the limits of conventional agrarian capitalism (including its supportive state structures and regulatory processes), even in the relatively rich and consumerist and environmentally conscious regions of Western Europe (see Marsden, 2018). Despite these trends, we are now also witnessing at a global and state levels (Paris COP21, Glasgow, COP26, and UN Sustainable Development Goals), and recent financial investment and disinvestment trends, the contested rise of post-carbonised transitions in the fields of energy, transport, infrastructure, and agri-food. These trends were not apparent at the onset of Kautsky's agrarian capitalism, nor were they seriously considered during the era when belief about inexhaustible unfettered spatialised growth of exploitable carbonism predominated.

Continuities

Given the significant differences in the structural conditions that have emerged since Kautsky, it might be tempting to dismiss his work as no longer relevant. We argue, however, that the current on-going transformations in the agri-food nexus and rural development suggest that his focus on how agrarian transformational processes occur remains conceptual valuable. This is principally because the onset of a post-carbonist and potentially post-capitalist regime in the 21st century, may refashion and substantially extend some of Kautsky's central tenets. In particular, we may now conceive the potential for land and biosphere holders (the 'new peasants'), and new consumer-based alliances, actively empowering themselves around these rural development nexus processes. Consequently, as Kautsky recognised, the recast ecological metabolism that is now (laterally) extending from new production to mass consumer action and practices on the one hand, and (vertically) from the biosphere and the lithosphere on the other, returns as an active and empowering force. This is partly a result of the increasingly necessary reliance in post-carbonist transitions in assembling a wider diversity of human and ecological resources, upon which energy, food, and material needs can be fulfilled.

Once we move away from an exploitative carbonised track of development, Kautsky's metabolic features become more relevant. These re-integrations are explored in greater detail below, but, in short, they are now being recognised in recent literature relating to: the recognition of the contemporary processes of re-peasantisation and de-peasantisation occurring at the same time (see Hebinck, 2018); the re-emergence of the metabolic significance of agri-food in urban as well as rural settings (see Friedmann, 2018), and the emergence of new and more distinctive material practices between food energy, water, and air for renewable energy and food production. Moreover, the binaries between capital, labour, and the peasantry have unfolded and become far more malleable and interactive (see van der Ploeg, (2017). Critical here, from a political ecology perspective, is a focus upon re-embedded power relations, not just those emerging from capitalists or the state, but those being actively re-assembled by variable groups of actors in more highly differentiated systems. These power relations transcend spatial and temporal scales (see Lamine et al., 2019; Rossi et al., 2019).

We will begin to deepen this analysis by considering two connected arenas. Firstly, what we call the emerging the agri-food/energy nexus, which, with the onset of post-carbonism and renewable energy production, once again becomes a more naturalised and land-based distributed activity, providing significant opportunities for land-based groups of farmers to develop more multi-functional and sustainable systems (see Figure 4.1). Secondly, we focus on more internalised transformations within agri-food systems and rural development, showing how political and social space can be found to carve-out new and reconfigured power relations in embedded place-based networks. In the following sections, we consider the shifting boundaries between common rights and private rights to land and the bio-sphere, and then, in more empirical detail, as to how these processes are playing out in the spatial and national context of the UK.

Land, shifting commons, and sustainability

A key place-based development dynamic represented in the above discussion, and in need of further and deeper consideration here, is the contested and shifting

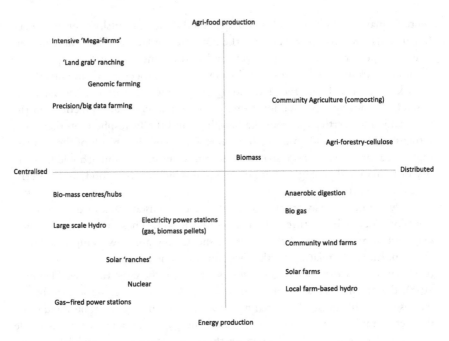

Agri-food production

Intensive 'Mega-farms'

'Land grab' ranching

Genomic farming

Precision/big data farming

Community Agriculture (composting)

Agri-forestry-cellulose

Biomass

Centralised — Distributed

Bio-mass centres/hubs

Anaerobic digestion

Bio gas

Large scale Hydro

Electricity power stations
(gas, biomass pellets)

Community wind farms

Solar 'ranches'

Solar farms

Nuclear

Local farm-based hydro

Gas–fired power stations

Energy production

FIGURE 4.1 The new multi-functional agri-food/energy nexus.

boundaries surrounding common-pool resources. Land, the wider biosphere, and lithosphere are a central part of this consideration. With the rise of post-carbonist processes especially in rural areas, the privatised and rentier capture of land, water, air, and sunlight, is becoming more manifest. Consequently, the social, legal, and governance boundaries between 'the commons' and private rights have become key areas for renewed rural development research. In the broader environmental field, the commons have already become an area of research interest. 'Commoning' is once again a central part of the new agrarian question, as it was for Marx and Kautsky. This is so given, as we shall see in the sections below, the countertendencies of further intensive rounds of privatised rentier rural based capitalism, whereby land rights are re-commodified for the benefits of landowners.

Enclosure and disempowerment are by no means a new phenomenon. In a widely cited passage of The Wealth of Nations, Adam Smith (1831) realised that when the reduction of common rights occurs, at least to an extreme extent, inequalities grow and 'morbid consequences' ensue:

> As soon as the land of any country has all become private property, the landlords, like all other men, love to reap where they never sowed, and demand a rent even for its natural produce. The wood of the forest, the grass of the field, and all the natural fruits of the earth, which, *when land was in common*, cost the labourer only the trouble of gathering them, come, even

to him, to have an additional price fixed upon them. He must then pay for the license to gather them: and must give up to the landlord a portion of what his labour either collects or produces. This portion constitutes the rent of land.

(our emphasis) (quoted in Christophers, (2019)) p. 57)

This is by no means an exclusively rural phenomenon, with rentier appropriation now rife across most of the world's cities, but it is important in re-considering the growth of rentier landholding (by the state and private capital) as a key process of re-appropriation and enclosure of rural land rights. This provides a conceptual, material, and social location for commoner struggle, not least by new social movements to recreate commoning and for small holders to protect against the monopolistic tendencies of rentier insurgency.

New enclosures in urban and rural Britain: restricting the public realm

Part of the wider process of growing neoliberalism in the UK since the 1980s has been the privatisation of former public land rights (see Christophers, 2019). This started with the explicit Thatcherite policy of selling council housing in the 1980s. In 1979, a third of all of Britain's housing stock was socially rented, a proportion only exceeded in Scotland, where up to 50 percent of the housing stock was socially rented. Devine noted that by the 1970s Scotland (and indeed Wales) had '...probably the largest share of public housing of any advanced economy outside the Communist bloc' (see Jack, 2019). Whilst the selling of council housing was a flagship ideological strategy, this was just the start of a wider process of the privatisation and financialisation of UK land in urban and rural settings. In 1979, 20 percent of the land was publicly owned, but subsequently, 2 million hectares (about 10 percent of the UK's landmass) have been privatised (see Christophers, 2019).

Much of the public land that has been sold has been of prime quality. Forests, national defense sites and municipally owned farms, for example, have been bought by financial institutions and housing development firms. Yet much of this land has been sold at significantly devalued rates. This has often been caused by a UK government financial squeeze on local authorities, who have therefore been forced to sell off assets at a devalued rate in order to attract private developers, in order to realise an immediate capital gain. The relaxing of public planning consent occurred in tandem with local authority financial constraints, thus allowing private sector development gain to proceed at a rapid rate. This has consequently increased property prices and caused major housing affordability issues for low-income groups in urban and rural communities.

The rise of the new *rentier class* whereby private landlords and a host of internationalised investors have sort to enclose through free-hold ownership of land

and property, whist then letting these assets to needy tenants, have ceased the opportunities presented by successive neoliberal governments. This process has intensified since the financial crisis of 2007–2008, and the austerity policies that followed. Land privatisation shows no sign of abating, especially since much of its regulation has been outsourced to private accountancy firms and land agents – the privatisation of privatisation. Unlike many other EU countries, the UK has effectively become a rentier economy, with the affordability of private rent becoming an increasing problem for many families. Rent has thus become a major source of economic growth and uneven wealth accumulation in the UK.

There are at least three important points to delineate from this process of enclosure:

- The new enclosure of the UK's former public commons has proceeded as largely a hidden process, with limited public scrutiny, and where even the administration of the privatising process has been effectively privatised. Moreover, registers of land ownership are now often shrouded in off-shore tax havens.
- Private developers have the freedom to hoard land and property assets until maximum market prices for selling or developing land can be achieved. Despite a major housing shortage, developers have the freedom to bank their land assets indefinitely. Such a tactic often increases the development gain between land with and without planning permission. This can be seen as the social and political construction of scarcity.
- There has been a continuous process of disempowerment, especially for the low-income urban residents and the young who have been excluded from any benefits of the devaluation of sales of council houses.

During the 18th and 19th centuries, the peasant or cottager having lost their cows, grazier rights, and rights to forage, were subsequently forced to purchase milk and food they needed (see Hammond and Hammond, 1913). In urban settings today, low-income groups and many of the young experience parallel hobsons' choices when faced with the eradication of former common rights. Forced to pay higher housing, energy, and food costs, and have less available green space, the re-domesticated urban poor are impoverished, whilst landlord's often extract excessive asset wealth. In the rural context, one feature of this process of the privatised enclosure has been the sale of former county council farm holdings. Since the First World War, and the onset of domestic food security concerns during the Lloyd George governments (see Marsden, 2017), local authorities across the UK bought land to make available for new agricultural small holders. Small family farms were encouraged to join the farming ladder and previously non-farming groups were given access to land. By 1926, around 1.5 percent of British land was owned by local governments and made available for agricultural use and occupancy. Since the late 1970s, the County Farms Estate halved from 426,000 acres in 1977, to 215,000 in 2019. Whilst there is a growing realisation that the processes of enclosure since the 1980s have been a significant factor in growing social and spatial inequality in the UK (see for example Christophers, 2019), connections between these processes and the environmental crisis have yet to be made.

There has been a more recent interest in the land question. Scotland, with its heightened political consciousness on the land question caused by the historical experiences of land clearances, has been most active on this issue. Community land rights have gained more recent interest in Wales (see Blake ., 2019), and this interest is being closely linked to environmental sustainability goals. The enclosure/empowerment dialectic is a vibrant and critical field in which to study place-based development processes. In a new era of global resource scarcity, and a crisis in neoliberalised systems of governance, we may be witnessing transformations in these balances and urban and rural communities are gaining greater consciousness and demanding that public bodies are more proactive. Consequently, the environmental and social crises are politically re-defining the need for a re-attribution of the public realm.

Conclusions: towards a place-based eco-transversalism?

In this chapter, we have argued that historical and contemporary struggles over nature are not a distinct arena or sectoral phenomena, but are deeply entangled with historically related material struggles over labour, care, land, race, class, gender, and the body. This is essentially a struggle over social nature, which involves the dialectics of property rights, and engages critical empowering and disempowering concepts such as domestication, enslavement, exposure, vulnerability, ambience, and milieu. Continuity in this process has been periodically highlighted, with the link between capital's assault on nature with enclosure, and class and place injustice. William Morris's eco-socialism, for example, included a powerful aesthetic dimension, which was adopted during the 20th century by conservationist and landscape movements in Europe and North America. Similarly, Frederick Engels highlighted the link between industrialism, public health, and social class and the co-evolving 'dialectics of nature'. More recently, Jason Moore (2017) reapplied these social ecologies, arguing that capitalism is running out of places and spaces in which it can colonise and continue to accumulate – the 'end of cheap natures'.

Such dialectics continue to be relevant in contemporary settings as vested interests, largely ignoring the continued concentration of power and unsustainability of agglomeration, continue to project a neoliberalised solution to the environmental and social crises. A green-capitalist imaginary has thus emerged, which incorporates an abstract (and non-place based) economising and commoditising logic based upon net gains, losses, and offsets. This abstraction assumes that nature can be parcelled into fungible, commensurable units whose place embeddedness and uniqueness, qualitative, intrinsic, and experiential qualities can be traded. Assigning a value to a natural asset according to how much various and competing actors would pay, to realise their competitive preferences assumes that net environmental gains can be re-created (see Watson 2021). This economistic and neoliberal logic is currently embodied in the global net-zero targets and visions, often completely detached from the sustainable values of indigenous communities associated with forest or fisheries stocks and flows.

Set against this background, Nancy Fraser (2021) argued for a connected 'anti-capitalist and transenvironmental eco-politics'. Departing from the historical environmentalism of the rich, she argues that many environmental justice movements '…are targeting entwinements of eco-damage with one or more axes of domination, especially gender, race, ethnicity and nationality, and some are explicitly anti-capitalist' (p. 125). Yet, she argues, the various fragmented movements such as de-growth, do not yet have a sufficiently substantive trans-environmentalist perspective. De-growth proponents, she suggests (see, for example Tim Jackson, 2021), tend to conflate what must grow in capitalism (namely value), with what should grow, but can't within capitalism (goods, relations, activities that can satisfy and sustain the vast expanse of unmet human needs). Stirling (2021) argued, 'what is missing to date is a clear and convincing perspective that connects all of our present woes, ecological or otherwise, to one and the same social system- and through that to one another' (p. 30). Our exploration of 'regenerative collectivism' in Chapter 10 is our limited attempt to contribute to this perspective.

Faith in future alliances being formed between related post-capitalist theses does appear to exist, but there are, we believe, several points of qualification needed regarding this faith in left eco-thinking:

- Often overlooked is the pragmatic and more prosaic dimension of place and space as a key organising and mobilising mechanism for conditioning the very differentiated ways in which progressive alliances may emerge.
- Pathways towards sustainable transformations need to start, develop and reach some level of development and attainment, but trajectories are not the same everywhere. Nor are the start and endpoints. Whilst much of the theory of transitions may assume rather deterministic outcomes, we must have more confidence to suggest pragmatic and open-ended transition pathways, with different outcomes in different places.
- Pragmatic and open-ended assumptions that incorporate place-based contingencies and different assemblages of powerful actors may develop in a co-evolutionary way alongside the crisis-ridden, but nevertheless still active

industrial capitalist and financialised/centralised system. In some places, the vibrancy and complexity of the former may reach a point when the latter becomes a marginal activity.

For us, rooted place-based action has the potential to accumulate into wider movements from below, with help from above. In this chapter, we have seen the centrality and historical continuity of struggles to emancipate people from environmental enslavement, which has been placed upon them, not least through a regressive process of property right enclosures. The environmental rights attributed to the urban worker, for instance, has been constantly squeezed, and their options as to what they metabolise, where they re-create, what type of water and air they use have been regressively constrained. Changing these circumstances requires integrating human ecologies with a wider vector of ecological practices and resources, de-coupled from monopoly capitalist logics. It requires us to redefine and re-interrogate places as emancipatory agents which create new freedoms.

References

Bai X, Nath I, Capon A, Hasan N and Jaron D (2012) 'Health and wellbeing in the changing urban environment', *Current Opinion in Environmental Sustainability*, 4(4): 465–470.

Blake, C (2019) Project Skyline: Report of Pilot Project. Cardiff Bay 2019 Project Skyline Report, Cardiff, Wales.

Christophers B (2019) *The New Enclosure: The Appropriation of Public Land in Neo-Liberal Britain*. London: Verso.

Davis M (2018) *Old Gods – New Enigmas: Marx's Lost Theory*. London: Verso.

Fraser N (2021) 'Climates of Capital: for a trans-environmental eco-socialism', *New Left Review*, 127(January/February 2021): 94–128.

Friedmann H and McMichael P (1989) 'Agriculture and the State System: the rise and decline of national agriculture', *Sociologia Ruralis*, 21: 93–117.

Friedmann H (2018) 'Metabolism of Global Cities: London, Manchester and Chicago', in: Chapter 66, pp. 1238–1359 in T K Marsden (ed.) *The Sage Handbook of Nature*. London and Los Angeles: Sage.

Gardiner B (2019) *Choked: Life and Breadth in the Age of Air Pollution*. Chicago, IL: University of Chicago Press.

Goodman D, Sorj B and Wilkinson J (1987) *From Farming to Biotechnology: A Theory of Agro-Industrial Development*. Oxford: Blackwell.

Greenberg P (2018) 'In the Valley of Fear', pp. 91–93 in: M Seaton (ed.) *New York Review of Books*, December 2018–January 2019. New York. Ny.books.com.

Hammond and Hammond (1913) *The Rural Labourer. Vol 1 and 2*. London: LSE Press, The Strand.

Harvey D (2021) 'Rate and mass: perspectives from the Grundrisse', *New Left Review*, 130(July/August 2021): 73–101.

Hebinck P (2018) 'De-/re-agrarianisation: global perspectives', *Journal of Rural Studies*, 61: 227–235.

Jack I (2019) 'Why did we not know. Review of Christophers op. cit', in: M Seaton (ed.) *London Review of Books*. 23 May 2019. Ny.books.com. New York.

Jackson, T (2021) *Post Growth: Life after Capitalism*. Cambridge: Polity.

Kropotkin P (1904) 'The desiccation of Eur-Asia', *Geographical Journal*, 23, 6, June 1904: 1–20.

Lamine C, Darnhofer I and Marsden T K (2019) 'What enables just sustainability transitions in agri-food systems? An exploration of conceptual approaches using international comparative case studies', Editorial of the Special Issue in *Journal of Rural Studies*, April 2019. 68: 1–7.

Lanchester J (2019) 'After the fall', *London Review of Books*, July 2018, 40(13): 1–5.

Latour B (2018) *Down to Earth: Politics in the New Climate Regime*. Cambridge: Polity Press.

Lee Y, Tan D, Siri J, Newell B, Gong Y, Proust K and Marsden T (2020) 'The role of public health dietary messages and guidelines in tackling overweight and obesity issues', *Malaysian Journal of Nutrition*, 26(1): 31–50.

Li Y, Westlund H and Liu Y (2019) 'Why some rural areas decline while others not: an overview of rural evolution in the world', *Journal of Rural Studies*, 68: 135–143.

Marsden T K (2017) *Agri-Food and Rural Development: Sustainable Place-Making*. London: Bloomsbury.

Marsden T K (ed.) (2018) *The Sage Handbook of Nature*. London, Los Angeles: Sage.

Marsden, T K and Rucinsca K (2019) After COP21: Contested transformations in the energy-food nexus. *Sustainability*, 11(1695): 2–17.

McCarthy J (2019) 'Authoritarianism, populism, and the environment: comparative experiences, insights and perspectives', *Annals of the American Association of Geographers*, 109(2): 301–313.

Moore J (2015) *Capitalism in the Web of Life: Ecology and the Accumulation of Capital*. London: Verso.

Morgan K, Marsden T K and Murdoch J (2006) *Worlds of Food*. Oxford: Oxford University Press.

Overseas Development Institute (ODI) (2016) *Population change on the rural developing world: making the transition*. ODI briefing paper, Keats S and Wiggins S, April 2016, London.

Purdy J (2015) *After Nature: A Politics of the Anthropocene*. Cambridge, MA: Harvard University Press.

Rasmussen L V (2018) 'Social-ecological outcomes of agricultural intensification', *Nature Sustainability*, June 2018, 34: 275–282.

Rossi A, Bui S and Marsden T K (2019) 'Redefining power relations in agro-food systems', *Journal of Rural Studies*, April 2019, 68: 147–158.

Scott J C (2017) *Against the Grain: A Deep History of the Earliest States*. New Haven, CT: Yale University Press.

Smith, A (1831) *The Wealth of Nations*. London: Inquiry.

Stirling A (2021) 'Preface: Branching pathways in agro-ecological transitions', pp. 23–33 in: C Lamine, D Magda, M Rivera-Ferre and T K Marsden (eds.) *Agro-Ecological Transitions, between Determinist and Open-Ended Visions*. Brussels: Peter Lang.

Strittmatter (2019) *We have been harmonised: Life in China's Surveillance State*. London: Old Street Publishing.

Thompson F M L (ed.) (1982) *Themes in Human History: The Rise of Suburbia.* Leicester: Leicester University Press and St Martin's Press.

Watson R (2021) *Making Peace with Nature: A Scientific Blueprint to Tackle the Climate, Biodiversity and Pollution Emergencies.* Nairobi: United Nations Environment Programme.

van der Ploeg J D (2017) Differentiation: old controversies, new insights. *Journal of Peasant Studies.* 45(3): 489–524.

5

THE CULTURAL PLACE

Culture is rarely considered a separate sphere in sustainable development. This might not necessarily be an omission, but rather a way of acknowledging the overarching role that culture plays. In fact, some identify culture as the 'the glue that binds together' everything else (Rana and Piracha, 2007, p. 21, citing Xian, 2000). Composed of 'attitudes, beliefs, mores, customs, values, and practices' (Rana and Piracha, 2007, p. 21), culture shapes our worldview and becomes the building block for identities, for ways of interrelating with the environment, and for political and economic behaviour. As such, culture has an intimate relationship with place. As we discuss in case studies in Chapters 8 and 9, local cultures significantly impact the world view and 'horizon' of communities and can strongly influence political and economic decisions.

For us, one of the essential cornerstones of our culture is food. Our relationship with food goes well beyond the instinctual consumption that keeps us alive, being intricately intertwined with our identities, heritages, religions, statuses, connections, memories, and values. Recognising this deep connection between food, culture and place, various food-related practices, skills, and product features in UNESCO's Intangible Cultural Heritage database. Some telling examples include, for instance, truffle hunting and extraction in Italy, which has been orally transmitted inter-generationally for centuries, through stories, fables, and anecdotes. The practice requires deep place-based knowledge, as well as tools and techniques that do not disturb soil conditions and that respect the annual regeneration of the species. Truffle hunting is also associated with community feasts, signalling the change of seasons, and showcasing the regional cuisine. Other similar UNESCO entries include the preparation of *Keşkek* – a traditional Turkish dish that involves the entire community in hulling the wheat, cooking, and eating. The entire process is accompanied by music and other performances, gathering people from across various ages and social statuses. Another example is the *Mediterranean*

DOI: 10.4324/9781003221555-6

diet – as the set of skills, knowledge, and traditions concerning growing, harvesting, fishing, cooking, sharing, and consuming food, in which women play a key role (UNESCO, n.d.). All these examples demonstrate how cultural diversity is shaped, and in turn shapes places, practices, and people, highlighting at the same time the connection between various Sustainable Development Goals (SDGs).

Certainly, food connects us with our families and our communities, and food practices are often passed down from generation to generation. When moving to another country, migrants often maintain the habit of cooking traditional or national foods as a way of preserving identities and staying rooted, while also constructing a sense of place. Their families might continue sending them national products, as an expression of care and to ensure a connection with the world left behind. Despite century-old cuisines, recipes, and products, food practices are dynamic, constantly evolving depending on the availability of products, trends, changing preferences, and often steered by marketing and the food industry. In this sense, food can also be a way of encountering and experiencing other cultures. The number of international restaurants in a place can be seen both as a sign of multiculturality and openness and as the trade-off for cultural losses or transformations. Studying cuisines, dishes and food practices can be a revealing indicator of how places are interconnected, serving as a testament to historical ties that surpass current national boundaries.

Furthermore, focusing on food can help us grasp some of the complex issues surrounding the place, culture, and sustainability. The globalisation of food has opened ways to explore other cultures, exchange products, and practices, and has enriched us in many ways. Concomitantly, it has also produced certain losses and has severely impacted the planet. Long supply chains, unequal power relationships in agriculture, trade, and consumption, as well as overproduction, overconsumption, and over-waste are all facets of globalisation and capitalism. Thus, this chapter explores questions and interlinkages around environmental damage, health, inequality, waste, and gender roles, highlighting how our worldviews are reflected in our food systems, and how both are currently on an unsustainable trajectory. Using the metaphor of food allows us to reflect on production, distribution, consumption, and waste management systems to delve into various issues that ultimately stem from our cultures and worldviews. Given the projected global population increase at least until 2070 when we might reach between 9.7 billion and 10.9 billion people (Adam, 2021), as well as corresponding continued economic growth and urbanisation, feeding 10 billion people will only worsen the current environmental issues under a business-as-usual scenario. Thus, changing our relationships with food could be one of the first and easiest steps we take in changing our environmentally damaging cultures.

Foundations: farming cultures

Food has often been used as a lens to understand human evolution. Yuval Noah Harari, for instance, presents the agricultural revolution as 'history's biggest fraud' – the moment when humans stopped living as hunter-gatherers in exchange for cultivating wheat (Harari, 2015, pp. 184–230). The historian explores how wheat cultivation seems to have brought few benefits for communities at that time: it offered lower quality nutrition compared to the variety of plants eaten before by hunter-gatherer tribes, it diminished food security, and its cultivation required significantly harder work to which the human body was not adapted. Most importantly though, it effectively reduced people's freedom, requiring them to settle and dedicate to agriculture. According to Harari, 'We did not domesticate wheat. It domesticated us' (Harari, 2015, p. 193). Nonetheless, wheat did bring humans a significant benefit. Its cultivation allowed homo sapiens to multiply exponentially, as it provided much more food per unit of territory compared to wild plants. Thus, wheat supported humans' evolutionary success (measured in the number of DNA copies), and permanently altered lifestyles and life quality.

Looking at the agricultural revolution from this perspective allows us to recognise a couple of characteristics that still hold true for our food systems: our reliance on a limited variety of plants, as well as the continuing predisposition for quantity over quality. Starting with the first of these – today, four crops (sugar cane, maize, wheat, and rice) account for 50 percent of global primary crop production (FAO, 2020). This has several consequences at both individual and wider levels. Diets are often not varied enough, as other local micronutrient-rich foods are replaced with these mainstream, globally traded crops. The farming of a limited number of crops, leads, eventually, not only to a loss of biodiversity and soil quality but also to reduced cultural diversity. This happens due to the embedded knowledge associated with specific plants – knowing how, when, and where to grow them, use, prepare, and consume (Carolan, 2018). Furthermore, as lands are mostly farmed as monocultures, they lead to soils that are degraded or depleted – one of the 'greatest ecosystem disservices that has resulted from the conversion of natural ecosystems to annual crop agriculture' (Crews et al., 2018). Besides, the production is dominated by a handful of countries: China, India, Brazil, the USA, Indonesia, and Russia. This hyper-concentration means that these exporting countries have a significant (positive and negative) impact on prices whenever harvests are affected (FAO, 2020). Furthermore, in its newest research findings, the International Land Coalition (2020) appreciates that more than 70 percent of the world's farmland is operated by the largest 1 percent of farms. This 1 percent is also the ones integrated into the corporate food system, exerting significant influence over the course of development of certain areas, and benefiting from economies of scale. As opposed to this, most farms that are smallholdings (80 percent) are steadily losing access to land, being squeezed into smaller parcels, or forced out altogether. Due to their limited size and power, small farms are either excluded from global food chains (International Land Coalition, 2020) or are losing their ability to choose what they grow and how (IPES FOOD, 2017).

Land inequality is, furthermore, intersectional in nature, as some of the most affected people include indigenous groups, women, youth, and elderly people. Considering specifically the question of gender shows that the gender gap in agriculture is far from closed. Although at the global level women represent 43 percent of the agricultural labour force (this figure is lower in Europe), they have significantly smaller chances to be a landowner, or a landholder compared to men. Besides, when women own land, they tend to own less than men do (FAO, 2018). This persistent issue, noticeable in many other sectors too, is reducing the chances of reaching the fifth SDG (Gender Equality), and probably impeding other goals. Research shows that when women have access to better incomes, as well as better education, child nutrition, health, and education tend to improve (FAO, 2011). Further gender-based issues will be explored in subsequent chapters, particularly Chapter 9 where our Deep Place study in Vanuatu explores these issues in rural and urban contexts.

Let us now consider the second abiding characteristic of the agricultural revolution – the human tendency to favour quantity over quality. While the global population has more than doubled in the past 60 years (from 3 billion in 1960 to almost 8 billion today), agricultural outputs have grown at an even faster rate, tripling (OECD, 2021). Today, the two are currently out of sync, as we are producing much more food than we need, whilst failing to eradicate hunger (as discussed in the following part). (Over)production puts a significant burden on the environment, and today's food supply chain accounts for 26 percent of anthropogenic GHG emissions, 32 percent of global terrestrial acidification, and 78 percent of eutrophication (Poore and Nemecek, 2018). Humanity's love for meat products plays a significant role here, as the proportion of agricultural land dedicated to livestock and their feed is more than 70 percent of the total (Dasgupta, 2021, p. 35).

The amount of food we produce – and as we are about to discuss, often waste – leaves a visible mark on places all over the world. Food production leads to water depletion, deforestation, and the so-called 'dead zones', is considered the most significant driver of terrestrial biodiversity loss (Dasgupta, 2021, p. 35). And yet, food overproduction is not an accident, with organisations such as OECD acknowledging that 'policies around the world tend to use highly distorting measures, often creating incentives for overproduction and overuse of inputs' (OECD, 2021). Sometimes, subsidies such as crop insurance and disaster relief are vital for producers facing climate change or sudden changes in the market (Bossio et al., 2021). Nonetheless, many of these subsidies are encouraging practices that are harmful to the environment, and ultimately to farmers, as they focus exclusively on production and output. Thus, as explored in detail in Chapter 2, the general fixation on perpetual growth is also a visible characteristic of food systems. This is particularly visible in countries across Southeast Asia, where growth in both urban populations and dietary changes have important implications for farming and rural food systems. For instance, the Malaysian government, faced with growing costs of imports of basic foods (e.g., meat, milk, and sugar), is promoting a domestic policy of self-sufficiency. This entails large-scale restricting of some small family farming areas, and the development of large-scale intensive livestock units. Subsidies are offered to high-tech intensive systems that are deemed suitable to feed growing urban populations. The intensive livestock industry is also highly dependent upon the procurement of global feedstocks, largely based on soy and corn. The reshaping of the Malaysian diet then becomes a major driver for 'locking in' intensive systems of enclosure both in Malaysia and abroad.

Besides public subsidies, research has linked overproduction to technological advances and cheap fossil fuels, showing that large-scale agri-food systems are not necessarily more resource-efficient than short and alternative food systems (Messner et al., 2020), as conventionally thought. In fact, there might be an inverse relationship between farm size and productivity due to 'a more (absolute) efficient use of land, water, biodiversity, and other agricultural resources by small farmers', especially for those nurturing polycultures (Carolan, 2018, p. 252). Still, flawed financing mechanisms are disguising the real cost of our food and the economic costs of the global food systems' associated negative impacts 'are credibly estimated to be higher than the entire market value of the global food system' (Bossio et al., 2021, p. 15).

Thus, as demonstrated here, food can be a useful lens to analyse and understand wider facets of global capitalism, from uniformisation, aspects of monopsony, and unequal terms of trade, to inequality and inefficient incentives which eventually lead to overproduction. Despite all this, food systems also hold great potential for positive change, as they are key to mitigate climate change and improve the livelihoods of producers' and consumers' health. We will return to this topic in the final part of this chapter. For now, let us advance through the food system, to explore another visible cost of our food surplus (apart from the environmental one), visible on human health. Consumption is estimated to exceed nutritional requirements by 20 percent (Hiç et al., 2016) and the growing trends of urbanisation might further worsen it. Thus, in the following section, we look more closely at the next stage of the food system, to explore the interrelated issues of health, distribution, cultural norms, and habits.

From farm to mouth...

Globally there are now more overweight and obese people, than underweight, and the number of associated deaths is following similar trends. For children and adolescents, the prevalence of overweight and obesity has more than quadrupled in 30 years, from 4 percent in 1975, to 18 percent in 2016 (World Health Organisation, 2021). To some, this phenomenon might not come as a surprise given the increase in food outputs previously discussed. Indeed, the world has been consuming ever-increasing portions and higher calorie intake coming primarily from vegetable oils, meat, and sugars. Still, while there are certain connections, these issues deserve a closer look. While it is true that overweight is caused by an imbalance between calories consumed and calories expended (World Health Organisation, 2021), the origins of the problem are not necessarily overconsumption. In fact, the increasing obesity rates are another facet of inequality, as more often than not, healthy and nutritious foods are inaccessible for economically deprived individuals and families. Although we spend less of our annual incomes on our food compared with previous generations, healthy diets are approximately five times more expensive than diets that only meet dietary energy needs through a starchy staple (FAO et al., 2020). Moreover, in reality, cheap foods are neither inexpensive nor desirable, as their cost does not reflect the full cost paid by societies (Carolan, 2018).

Besides the negative environmental impact explored in the previous section, cheap foods limit our abilities, destabilise our cultural systems, and greatly affect human health and well-being, not to mention animal welfare. These issues are most visible in many low- and middle-income countries, which are now facing the so-called 'double burden of malnutrition' (World Health Organisation, 2021): while they continue to struggle with undernutrition, they are also experiencing an upsurge in obesity. Obesity is recognised not only as a disease in itself but also as a condition that increases the likelihood of developing a range of non-communicable diseases (NCDs; include diabetes, heart disease, stroke) and mental health illnesses (World Obesity Federation, 2021). Besides, although we clearly need more time to fully understand the longer-term implications of the COVID-19 pandemic, the initial evidence suggests that overweight and obesity, together with inadequate physical activity levels, can be correlated with hospitalisation cases (Hamer et al., 2020). Thus, obesity and its associated conditions have serious implications for healthcare systems and economies.

In Malaysia for instance, about 8 percent of total mortality each year is attributed to obesity. Beyond the increased risk of obesity-related chronic diseases and poorer quality of life, the healthcare costs of treating obesity-related disease conditions are rapidly escalating. On average, obese Malaysian males and females lose between 6–11 years and 7–12 years of productive life, respectively (Tan and Siri, 2018) Physical, social, and cultural environments associated with work, food, family, and community (Yoon and Kwon, 2014) all enable and constrain the individual choices and behaviours that affect obesity. For example, in Malaysia, the widespread practice of serving sweet and savoury snacks in morning and afternoon tea at functions, conferences, and meetings enables over-consumption of food and cements frequent eating as a social norm. Working hours, availability of fast food, and school nutrition (Tan and Siri, 2018), among other factors, also play key enabling/constraining roles in Malaysia. The popularity of fast food and the rapid expansion of outlets has especially affected food choices of Malaysian youth. On the one hand, it reflects a desire for convenience and a reduced appreciation of traditional food choices. On the other, accessible, and affordable fast food increases habituation to unhealthy diets. This is reinforced by fast-food advertising, commercials, and television programming aimed at children, fuelling addiction.

In Europe too, more than half of adults are overweight (36 percent pre-obese and 17 percent obese). This proportion tends to fall as the educational level (often positively correlated with the income level) rises, especially for women (EUROSTAT, 2021). Interestingly, in all European countries, men suffer from being overweight more than women, with gaps as big as 21 percent in Luxembourg (59 percent men versus 38 percent women). This is atypical and hard to explain, since worldwide, the female gender is considered a potential risk for developing obesity due to factors such as pregnancy and post-partum, hormonal treatments, different biological cravings for food, as well as socio-economic factors including disparate socio-economic statuses, dissimilar literacy rates, divergent customs, and social expectations (Kapoor et al., 2021).

In Wales, statistical analyses have shown that over 20 percent of all people living in the Cardiff Capital Region are obese, and 60 percent are overweight (Axinte, 2017). Combined with low levels of physical activity, considerable amounts of alcohol consumption, and smoking, these issues show that unhealthy lifestyles are very common both among young and adult inhabitants and that so-called developed nations are not necessarily doing better than developing ones. This situation (characteristic of the entire United Kingdom) is among the worst across OECD countries (OECD, 2017) and it underscores a variety of societal problems: from mental health issues to obesogenic environments and food deserts. An independent report showed that there are at least nine areas in Wales that can be considered 'deprived food deserts' – areas where there are less than two supermarkets or convenience stores, and which are part of the 25 percent most deprived areas, according to the Index of Multiple Deprivation (Corfe, 2018). Food deserts are places where fresh and healthy food is scarce, while high-caloric, non-nutritious options proliferate. They affect communities that have poor public transport connections and limited access to cars, and they rely on the corner shop for their daily food supplies. As seen throughout this book, these pockets of deprivation often co-exist in close proximity to rich areas, and understanding the difference in the lived experience can be advanced by the kind of Deep Place analysis we have undertaken in the communities detailed in Part Two. Various studies have shown that in the UK a boy born in an affluent area can live 8.4 more years, compared to a boy born in a poor area (Corfe, 2018). Poor diet is a contributor to the widening life expectancy gap.

Thus, despite producing more than enough food to feed every living person, the world is riddled by food insecurity. Food insecurity means that the possibility of obtaining food is uncertain, and people are forced to compromise on food quality, quantity, or both. The number of people affected by moderate and severe food insecurity globally has been slowly rising since 2014, and the COVID-19 pandemic will worsen the situation for at least 80 million people more (FAO et al., 2020). Approximately 2 billion people – a quarter of the world's population, are currently affected by food insecurity. All this shows that the world is not progressing towards achieving 'zero hunger by 2030', the United Nations' second SDG, neither in terms of securing universal access to safe and nutritious food nor in terms of ending all forms of malnutrition.

Beyond nutrition, health, and impacts on life expectancy, food can also reveal deeply engrained cultural and/or place-based characteristics. Certain foods have evolved as trademarks of specific areas, leading to the creation of labels of origin for food. Such an example is the EU's quality schemes, with three main types of geographical indication (GI): protected designation of origin (certifying the strongest links to the place in which products are made), protected GI (certifying that at least one of the stages of production, processing or preparation takes place in the region), and GI of spirit drinks and aromatised wines (European Commission, n.d.). This certification recognises that local factors – location, materials, and ingredients, together with local knowledge, factors of production, and preparation lead to unique products that are impossible to replicate elsewhere. These trademarks have various functions: to certify and protect quality products from imitation, to offer customers reliable information, and to support (typically) rural areas that depend on food production (Spognardi et al., 2020). While potentially benefitting Europeans, the EU GI schemes have been criticised for being protectionist, limiting innovation and fair competition at the global level (William Watson, 2016).

Despite these critiques, the idea of terroir – the connection between flavour and local environment – has the potential to reduce the negative impacts of globalisation and uniformisation without sacrificing openness and multiculturality. In fact, many of us learn about a country or a region through their food products. Think of feta, Gouda or Roquefort cheese, Prosciutto di Parma ham, Jersey Royal potatoes, Champagne, or Port wine for Europe, as well as Ethiopian coffee, Turkish delight (found outside of Turkey too, in various Arabic countries), or Sichuan peppercorns. Tasting a country's food can be the first encounter with a new world, a change of perspective that opens new horizons and potentially reduces the stigma attached to (allegedly) less-developed nations.

Besides the strong place nexus, food can also be a means to explore social and family relations. For instance, in various parts of Eastern Europe grandmothers are notorious for showing their affection not by saying 'I love you', but by asking 'Are you hungry?'. On the other side of the planet, the Chinese take it a step further using the question 'Have you eaten?' as the equivalent of a greeting. This demonstrates 'the key role played by food and its consumption in the creation of a social relationship, and hence, in society itself' (Ciorra and Rosati, 2015, p. 10). Indeed, more often than not, food surpasses its purpose of nutritional necessity. Events, both religious and non-religious ones, are often organised around meals, with elaborate dishes and long-established food pairings and sequences. For instance, a traditional Romanian wedding will provide guests with a cold starter (a plate combining cheeses, cold cuts, various cold salads, and spreads), a warm starter (often a meat soup or a fish-based meal), a first course (meat, generally beef or pork, and potatoes), a second course (meat-filled cabbage rolls with polenta), and cake. Fruits, pretzels, peanuts, and sometimes other types of desserts (more recently in the form of a cake bar), along with various types of alcoholic and non-alcoholic drinks will be always available. This menu is served in the span of 12 hours, making it virtually impossible to avoid food wastage. In many families, similarly, lavish celebrations are repeated on a yearly basis during major religious events such as Christmas and Easter. These practices are shaped by a mix of traditions, values, and social and religious customs, maintaining strong meanings in the individual and collective consciousness. In Britain, in particular, the centrality of the Turkey or Goose at Christmas owes much to the stories of Charles Dickens, and so food culture also reflects literature.

Shifting towards more balanced ways of celebrating can be extremely difficult as practices are passed from generation to generation and are maintained by certain societal expectations. Excesses are damaging not only due to the unnecessarily large quantities of food produced and consumed but also due to the heavy reliance on animal-based products. Since the 1950s, with the end of the 'unintentional vegetarianism' experienced by the world's poor (Motta, 2021), there has been a rise in mass meat consumption. In the Eastern Bloc (of which Romania was part of), this trend started growing around 40 years later, with the change of regimes. Globally, increasing average individual incomes and population growth led to an increase in both the global average per capita and the total amount of meat consumed. Meat-based diets are not only harmful to individuals' health (leading to a higher risk of certain types of cancer and other diet-related NCDs) (FAO et al., 2020) but have significant negative effects on water consumption, land use, and greenhouse gas emissions. Research has shown that 'meat, aquaculture, eggs, and dairy use approximately 83% of the world's farmland and contribute 56% to 58% of food's different emissions, despite providing only 37% of our protein and 18% of our calories' (Poore and Nemecek, 2018). These findings are showing that what we eat matters more than where our food comes from. In this sense, while actions to shorten food chains can reduce transport-associated emissions, policies and programmes that support the nutritional transition towards a plant-centred diet are instrumental. This is easier said than done, since in many parts of the world, meat and processed foods can be 'symbols of upward social mobility' (Motta, 2021), or a concentrated nutrient source for low-income families (Godfray et al., 2018). Attempts to tame or nudge consumption into more sustainable practices have often been opposed and condemned. A recent example took place in Spain, when the consumer affairs minister published a video encouraging Spanish citizens to reduce the amount of meat they eat, for their health and the health of the planet (Garzón, 2021). This brought outrage among some citizens, meat-producing associations, as well as other ministers, who saw the campaign as an assault on Spanish culture and on the work of Spanish farmers (Landauro and Allen, 2021).

Despite the scientific consensus on the need to change dietary habits and reduce meat consumption, leaders across the world are failing to transmit this message. A possible explanation for this failure might be the tension between personal freedoms, inscribed in law in many parts of the world, and sustainability. As Heidenreich (2018) explained, democratic regimes protect individual rights, including freedom of expression and choice of lifestyle, and are generally governed through collective choice. Nonetheless, 'maximized individual freedom is incompatible with the need to lower consumption', and sustainability transitions require significant changes (and often limitations) in the way people currently live, move, and consume (Heidenreich, 2018, p. 358). Besides, the right to make collective choices does not necessarily lead to sustainable ones, as we have seen in previous riots against fossil fuel taxes (e.g., gilets jaunes in France) or the 2016 US election (which eventually led to the country's withdrawal from the Paris agreement). The difficulties in communicating science and combating misinformation also highlight a certain friction between scientific knowledge and evidence-based decision-making versus allowing a maximum range of opinions, an ideal of deliberative democracy.

Furthermore, governments find it very difficult to maintain the balance between different goals, especially as some might be contradicting each other (e.g., SDG 8: Economy Growth and SDG 12: Responsible Production and Consumption). Even the promotion of *responsible production and consumption* is difficult to achieve, in a century defined by consumerism. Finally, sustainability requires long-term visions and strategies, which are incompatible with political short-termism and frequent changes of government and leadership. These tensions open the space for an important question: what would democratic cultures for sustainable place-making look like? The last part of this chapter returns to this question. For now, it is safe to say that food-related social practices are a constraint for behaviour change. As a result of structural overproduction and overconsumption (see Messner et al., 2020) explored so far, our societies are also characterised by mass food waste. The following section explores how, why, and where food waste happens, showing the intricate connections with the rest of the food chains, and inevitably, with cultures.

...or elsewhere, often left to rot

It is widely accepted that food waste is a major global issue, despite highly inconsistent measurements and terminologies. Cloke (2020, p. 66) uses Bourdieu's idea of doxa to deconstruct waste as something so generally accepted that it 'goes without saying because it comes without saying'. Nonetheless, he argues that:

> Waste needs to be explicated as a form of biopolitical power, as a social relationship rendering it both co-constitutive of and interdependent with value – food (indeed all consumer products) in this view is more than just a vehicle for waste – the potential for waste in food products and the waste involved in producing them is part of their value.
>
> *(Cloke, 2020, pp. 66–67)*

Before delving further into the waste paradox, it is worth exploring the scale of our (food) waste.

This is not, however, a straightforward endeavour. FAO's attempt to encompass the totality of food discarded across the entire value chain under 'food loss and waste' does not include

> crops lost before harvest because of pests and diseases, crops left in the field, crops lost due to poor harvesting techniques or sharp price drops, or food

that was not produced because of a lack of adequate agricultural inputs, including labour availability.

(Delgado et al., 2021, p. 3)

Even with this omission, older estimates led to the often-cited adage that 'if food wastage were a country, it would be the third largest emitting country in the world', right after China and the USA, with a total carbon footprint of 4.4Gt CO_2 equivalent per year (FAO, 2015). To put it in perspective, food-waste emissions are almost as high as global road transport emissions. Losses happen across commodities and in all stages of the food chain, resulting in roughly one-third of the food produced globally being discarded each year. Quantity-wise, vegetables, cereals, and starchy roots go to waste most often, whereas carbon footprint-wise the top three are cereals, vegetables, and meat. Food chain stages have different carbon intensities, and the further along the chain the food loss occurs, the more greenhouse gases are accumulated during harvesting, processing, distribution, and so on.

Nonetheless, the biggest quantity of food waste happens during agricultural production. Research from the WWF (2021) brings new insights into this issue, estimating that of all the food grown globally, approximately 40 percent goes un-eaten (a higher estimate than FAO's). The WWF findings are ground-breaking as they are also showing that food waste on farms is not only a problem in less industrialised countries, as previously thought due to the lack of technology (e.g., cooling facilities) or to access and transport. On the contrary, despite higher lev-els of mechanisation, high- and middle-income countries in Europe, industrial-ised Asia, and North America are responsible for 58 percent of the global harvest stage waste. WWF's study confirms that FAO's definition of 'food loss and waste' and United Nations' SDGs might not be able to capture the complexity of food waste. SDG 12.3 sets a 50 percent reduction target for retail and consumer food waste and mentions the need to reduce 'food losses along production and sup-ply chains, including post-harvest losses' (FAO, n.d.). Food surplus is sometimes diverted into the feed system or used as biofuels, and therefore excluded from losses. This has various repercussions, including the disincentive to address the roots of overproduction, as well as the undermining of food security and nutri-tion as primary objectives.

Furthermore, without a holistic understanding and precise measurement of the farm-stage food loss, we cannot reach SDG 12 that asks for responsible pro-duction and consumption. Even in the so-called developed countries, such as the UK, there is an acute lack of data on food waste pre-farm gate. Using data from other comparable regions and studies on a limited number of crops, WRAP (2021, p. 4) suggests that 'more food waste arises from primary production than from hospitality & food service and retail combined.' Primary production waste amounts to approximately 3.6 million tonnes per year, slightly less than house-hold waste (4.5 million tonnes). Although the retail and manufacture stages seem to be performing much better, they play an essential role in the production of waste. This happens due to strict (and sometimes ridiculous) aesthetics, shape, and size standards required from farmers, forecast orders submitted to suppliers in advance, but which might be different than actual purchases (placing the cost of overproduction on suppliers), as well as marketing campaigns and promotional offers that encourage overspending (Carolan, 2018).

Regardless of the stage though, through the food we waste, we not only emit significant amounts of greenhouse gas emissions into the atmosphere, but we practically also discard all the water, soil, fertilisers, and energy that went into the agricultural production. Indeed, as Messner et al. (2020) argued, there is a 'prevention paradox' in our food systems. Although it is commonly accepted that prevention is the best option, most governmental, NGO, and industry initiatives to reduce waste focus on waste once it already exists, rather than on averting it in the first place. These waste management strategies can be clustered into three main categories: (i) avoiding that surplus food is wasted, (ii) raising awareness among consumers to avoid throwing away food, and (iii) diverting food from going into landfill, through zero waste or circular economy ideas. While they should not be entirely discredited, such actions deflect us from the elephant in the room: overproduction, its environmental footprint, and the practices which lead to food waste (not least in terms of aggressive marketing, promotional offers, or increasing portion sizes). Our food systems are locked in waste generation due to the governmental incentives for growth (as discussed in the first part of this chapter) and legal regimes. For instance, waste prevention could violate international agreements and laws (Messner et al., 2020), while also potentially affecting farmers if not accompanied by wider programmes to support them. Besides, food waste is also locked in by the same physical infrastructure and investments dedicated to waste management.

Food waste consequently exemplifies much of the waste created in the wider manufacturing and consumption practices of globalised neoliberal capitalism. As it is obvious by now, hunger, malnutrition, and food insecurity are not due to insufficient amounts of food, as we could currently feed approximately 10 billion people. The causes for this mismatch include resource misallocation, mismanagement, vested interests, and entire economic systems depending on waste, as well as cultural values around food. Yet, food is only one of the many things we are used to throwing away. Waste generation has been increasing between 2010 and 2018 by 5–7 percent in the EU-27 (European Environment Agency, 2021). Globally, under a business-as-usual scenario, the World Bank estimates that the total quantity of solid waste generated will grow from 2 billion tonnes in 2016, to 3.4 billion tonnes by 2050. As waste generation levels are correlated to income levels and urbanisation, low-income countries are expected to see a double or even triple output (Kaza et al., 2018). The decoupling of economic growth and waste, and implicitly CO_2 emissions, is not happening, and the global economy is not becoming more circular either. Cloke (2020, p. 65) coined the term vastogenesis to encompass the fact that waste generation is a core part of mass consumption, and not an 'unfortunate side-effect of consumer capitalism'. Ever-increasing amounts of waste are lucrative and are currently being operationalised by various businesses and corporate actors, leading to an even more profitable consumer culture. Thus, any attempt at reducing waste is futile unless actions focus on addressing its systemic causation. Following this exploration of our food systems, patterns of production, consumption, and waste, and the socio-cultural, economic, and environmental connections, the final part of this chapter looks for signs of hope.

Changing our relationship(s) with food

This chapter has used the metaphor of food to discuss cultural(?) practices, norms, values, and habits, and their junctures with issues of environmental degradation, health, and inequality. Whereas many arguments refer to global phenomena, place-based examples were given wherever possible. The topics and places touched upon are by no means exhaustive, but they offer a departure point to examine and re-evaluate our relationship(s) with food. As the following paragraphs aim to conclude, this could be a first step towards changing our anthropocentric worldviews that commodify and exploit nature, depleting planetary resources at an unsustainable pace.

We have reached a point where sustainability is no longer enough. Sustaining has often translated into 'determining how much damage can feasibly be inflicted' (du Plessis, 2011, p. 7), and we have repeatedly failed to stay within the limits we set. As seen in the previous parts, global goals such as SDG 12 are not comprehensive enough to tackle some of our most urgent crises, like farm-level food waste. Indeed, we have been treating agriculture and 'the farm like an isolated, industrial machine', ignoring that the 'system of a farm relies on interactions with the larger natural system', which we cannot control (Ellen MacArthur Foundation, 2019). As a departure from this mindset, regenerative development advances from minimal or neutral environmental impact to creating positive effects for a 'mutually supportive symbiosis' between the built, cultural, and natural environments. Having degraded not only the environment but also our social and cultural systems, we need to do much more to restore and enhance ecosystems and community health (Wahl, 2017). Thus, considering the multiple crises that societies go through today, it is no longer sufficient to sustain or remain 100 percent neutral.

The regenerative paradigm was inspired by permaculture – an approach to agriculture that tries to work with nature, rather than against it, using systems thinking and nature-based solutions. Today, regenerative agriculture is gaining increasing interest from researchers and policymakers (see, for instance, IPCC, 2019), despite lacking certification (as organic agriculture does) or a commonly agreed set of practices that could protect it from greenwashing. Nonetheless, regenerative practices promise to recover soil health, capture carbon, restore degraded habitats, reduce chemical use and pesticides, and improve the overall connection with nature, and between communities. Unfortunately, this 'latent potential' has had limited power to transform the global food system, which is a product of its hard to transcend history (Bossio et al., 2021, p. 92), locked in an unsustainable regime. Important elements that shape our food systems material-ised at a time when climate change and biodiversity loss were not a priority, and their legacy is still visible in policies, tax systems, and practices.

Nonetheless, regenerative agriculture is only one of the many initiatives that have coalesced around food. Many others have gathered enough momentum to become social movements in their own rights, demonstrating that food can be a means to reach a bigger end. Although space does not allow for a detailed sum-mary here, it is worth mentioning Motta's (2021) work which demonstrated that food inequalities are multidimensional, multi-scalar, and intersectional in nature. As a response, alternative local food initiatives aim to reduce the environmental impacts of globalised food relations; food sovereignty movements react to class, racial, and gender inequalities and power asymmetries; food injustice movements fight against institutional racism; feminist movements tackle ongoing gender in-equality visible across the entire food chain; and vegan movements speak for interspecies justice, animal rights and environmental protections. While some might be present globally, these food initiatives are mapped onto different parts of the world, spanning both the Global North and the Global South. Unfortu-nately, as seen throughout this chapter, our food systems are characterised by high levels of cultural and social 'lock-in' which leads to automatic reproduction of unsustainable and unhealthy practices. For us, as we demonstrate through the case studies in Part Two, place-based change offers an important sphere within which we may start to overcome these practices.

Within place-based development, we see food as a core element, due to its centrality at both micro and macro levels. Beyond the role played for individual sustenance and health, food embodies memories and traditions, local and gen-erationally transferred practices, as well as symbols and statuses characteristic of communities, places, and specific periods. Communities (both small and large) can build their identities around food-related issues, and food can be a key ele-ment to develop a worldview supportive of regenerative development. Nurturing sustainable food production and consumption will be vital in the years to come, as approaches that restore the health of ecosystems will be needed to mitigate the effects of climate change and reduce the burden of malnutrition. As seen throughout this chapter, food systems are deeply interconnected with the bio-sphere, depending on natural resources, and often resulting in damaging outputs that periclitate food security for many areas and communities around the world. Although changes are not straightforward due to significant political, legisla-tive, and behavioural lock-ins, the local, place-based level is the most appropriate starting point.

References

Adam D (2021) 'How far will global population rise? Researchers can't agree', *Nature*, 597: 462–465. https://doi.org/10.1038/D41586-021-02522-6

Axinte L (2017) *Cardiff Capital Region youth profile*. Retrieved December 22, 2021 from https://www.cardiff.ac.uk/__data/assets/pdf_file/0011/776333/CCR_Youth_Profile-Lorena_Axinte-1.pdf

Bossio D, Obersteiner M, Wironen M, Jung M, Wood S, Folberth C, Boucher T, Alleway H, Simons R, Bucien K, Dowell L, Cleary D and Jones R (2021) *Foodscapes: Toward Food System Transition. The Nature Conservancy, International Institute for Applied Systems Analysis, and SYSTEMIQ*, ISBN: 978-0-578-31122-7

Carolan M (2018) *The Real Cost of Cheap Food*. New York: Routledge.

Ciorra P and Rosati A (2015) *Food dal cucchiaio al mondo*. Quodlibet.

Cloke J (2020) 'Interrogating waste: Vastogenic regimes in the 21st century', *Routledge Handbook of Food Waste*, 65–78. https://doi.org/10.4324/9780429462795-6

Corfe S (2018) *What Are the Barriers to Eating Healthily in the UK?* London: The Social Market Foundation, ISBN: 978-1-910683-51-4.

Crews T E, Carton W and Olsson L (2018) 'Is the future of agriculture perennial? Imperatives and opportunities to reinvent agriculture by shifting from annual monocultures to perennial polycultures', *Global Sustainability*, 1. https://doi.org/10.1017/SUS.2018.11

Dasgupta P (2021) *The Economics of Biodiversity: The Dasgupta Review – Abridged Version*. London: HM Treasury.

Delgado L, Schuster M and Torero M (2021) 'Quantity and quality food losses across the value chain: a comparative analysis', *Food Policy*, 98: 101958. https://doi.org/10.1016/J.FOODPOL.2020.101958

du Plessis C (2011) 'Towards a regenerative paradigm for the built environment', *Building Research & Information*, 40: 7–22. https://doi.org/10.1080/09613218.2012.628548

Ellen MacArthur Foundation (2019) *Regenerative agriculture* [WWW Document]. Accessed 22 December 2021: https://ellenmacarthurfoundation.org/articles/regenerative-agriculture.

European Commission (n.d.) *Quality schemes explained* [WWW Document]. Accessed 12 December 2021: https://ec.europa.eu/info/food-farming-fisheries/food-safety-and-quality/certification/quality-labels/quality-schemes-explained_en.

European Environment Agency (2021). *Waste generation and decoupling in Europe* [WWW Document]. Accessed 20 December 2021: https://www.eea.europa.eu/ims/waste-generation-and-decoupling-in-europe.

EUROSTAT (2021) *Overweight and obesity – BMI statistics* [WWW Document]. Accessed 28 November 2021: https://ec.europa.eu/eurostat/statistics-explained/index.php?title=Overweight_and_obesity_-_BMI_statistics#Education_level_and_overweight.

FAO (2011) *Women in Agriculture: Closing the Gender Gap for Development*.

FAO (2015) *Food wastage footprint & climate change* [WWW Document]. Retrieved December 20, 2021, from https://www.fao.org/documents/card/en/c/7338e109-45e8-42da-92f3-ceb8d92002b0/.

FAO (2018) *The Gender Gap in Land Rights*.Rome: FAO/IFPRI.

FAO (2020) *Statistical Yearbook 2020, World Food and Agriculture – Statistical Yearbook 2020*. https://doi.org/10.4060/CB1329EN

FAO (n.d.) *Metadata of Indicator 12.3.1 Global Food Loss Index*.

FAO, IFAD, UNICEF, WFP, and WHO (2020) *The state of food security and nutrition in the world 2020. Transforming food systems for affordable healthy diets*. https://doi.org/10.4060/ca9692en

Garzón A (2021) Twitter [WWW Document]. Accessed 16 December 2021: https://twitter.com/agarzon/status/1412715352325246990?s=20.

Godfray H C J, Aveyard P, Garnett T, Hall J W, Key T J, Lorimer J, Pierrehumbert R T, Scarborough P, Springmann M and Jebb S A (2018) 'Meat consumption, health, and the environment', *Science* (New York), 361. https://doi.org/10.1126/SCIENCE.AAM5324

Hamer M, Gale C R, Kivimäki M and Batty G D (2020) [Overweight, obesity, and risk of hospitalization for COVID-19: a community-based cohort study of adults in the United Kingdom', *Proceedings of the National Academy of Sciences of the United States of America*, 117: 21011–21013. https://doi.org/10.1073/PNAS.2011086117

Harari Y N (2015) *Sapiens: A Brief History of Humankind*. New York: Harper.

Heidenreich F (2018) 'How will sustainability transform democracy? Reflections on an important dimension of transformation sciences', *GAIA*, 27: 357–362. https://doi.org/10.14512/GAIA.27.4.7

Hiç C, Pradhan P, Rybski D and Kropp J P (2016) 'Food surplus and its climate burdens', *Environmental Science and Technology*, 50: 4269–4277. https://doi.org/10.1021/ACS.EST.5B05088/SUPPL_FILE/ES5B05088_SI_001.PDF

International Land Coalition (2020) *Uneven Ground. Land Inequality at the Heart of Unequal Societies*. Rome: International Land Coalition Secretariat.

IPCC (2019) *Climate Change and Land. An IPCC Special Report on climate change, desertification, land degradation, sustainable land management, food security, and greenhouse gas fluxes in terrestrial ecosystems* [WWW Document]. URL (Accessed 22 December 2021).

IPES FOOD (2017) *Too Big to Feed: Exploring the Impacts of Mega-Mergers, Consolidation and Concentration of Power in the Agri-Food Sector*. Brussels: IPES Food.

Kapoor N, Arora S, and Kalra S (2021) 'Gender disparities in people living with obesity – an unchartered territory', *Journal of Mid-Life Health*, 12: 103. https://doi.org/10.4103/JMH.JMH_48_21

Kaza S, Yao L C, Bhada-Tata P and van Woerden F (2018) *What a Waste 2.0: A Global Snapshot of Solid Waste Management to 2050*. Washington, DC: World Bank. https://doi.org/10.1596/978-1-4648-1329-0

Landauro I and Allen N (2021) *Pork-barrel politics? Spanish ministers clash over less-meat campaign* [WWW Document]. Reuters. Accessed 16 December 2021: https://www.reuters.com/world/europe/pork-barrel-politics-spanish-ministers-clash-over-less-meat-campaign-2021-07-08/.

Messner R, Richards C and Johnson H (2020) 'The "Prevention Paradox": food waste prevention and the quandary of systemic surplus production', *Agriculture and Human Values* 37: 805–817. https://doi.org/10.1007/s10460-019-10014-7

Motta R (2021) 'Social movements as agents of change: fighting intersectional food inequalities, building food as webs of life', *The Sociological Review*, 69: 603–625. https://doi.org/10.1177/00380261211009061

OECD (2017) *Obesity update* [WWW Document]. Accessed 28 November 2021: https://www.oecd.org/health/obesity-update.htm.

OECD (2021) *Making better policies for food systems*. OECD. https://doi.org/10.1787/DDFBA4DE-EN

Poore J and Nemecek T (2018) 'Reducing food's environmental impacts through producers and consumers', *Science*, 360: 987–992. https://doi.org/10.1126/SCIENCE.AAQ0216/SUPPL_FILE/AAQ0216_DATAS2.XLS

Rana R S J B and Piracha A L (2007) 'Cultural frameworks', pp. 13–50 in: M Nadarajah, A Tomoko Yamamoto (eds.) *Urban Crisis: Culture and the Sustainability of Cities*. Tokyo: United Nations University Press.

Spognardi S, Vistocco D, Cappelli L and Papetti P (2020) 'Impact of organic and "protected designation of origin" labels in the perception of olive oil sensory quality'. *British Food Journal*, 123, 2641–2669. https://doi.org/10.1108/BFJ-07-2020-0596/FULL/PDF

Tan D and Siri J (2018) (eds.) *SCHEMA Case Studies: Applying Systems Thinking to Urban Health and Well-Being*. Kuala Lumpur, Malaysia: United Nations University for Global Health.

UNESCO (n.d.) *UNESCO's interactive visual Living Heritage and Sustainable Development* [WWW Document]. Accessed 1 March 2022: https://ich.unesco.org/dive/sdg/.

Wahl D C (2017) *Sustainability is not enough: we need regenerative cultures* [WWW Document]. Accessed 22 December 2021: https://designforsustainability.medium.com/sustainability-is-not-enough-we-need-regenerative-cultures-4abb3c78e68b.

William Watson K (2016) *Reign of terroir: how to resist Europe's efforts to control common food names as geographical indications* [WWW Document]. Cato Institute. Policy Analysis No. 787. Accessed 12 December 2021: https://www.cato.org/policy-analysis/reign-terroir-how-resist-europes-efforts-control-common-food-names-geographical#not-intellectual-property-nbsp.

World Health Organisation (2021) *Obesity and overweight* [WWW Document]. Accessed 28 November 2021: https://www.who.int/news-room/fact-sheets/detail/obesity-and-overweight.

World Obesity Federation (2021) *COVID-19 and Obesity: The 2021 Atlas. The Cost of Not Addressing the Global Obesity Crisis*.London: World Obesity Federation.

WRAP (2021) *Food Surplus and Waste in the UK – Key Facts*. Cardiff: WRAP.

WWF (2021) *Driven to Waste: The Global Impact of Food Loss and Waste on Farms*. Woking: WWF-UK.

Xian, G (2000) Culture and development: A sustainable world in the twenty-first century, *Culturelink*, Special Issue: 173–176.

Yoon N H and Kwon S (2014) 'The effects of community environmental factors on obesity among Korean adults. A multi-level analysis', *Epidemiology and Health*, December 2014: 1–10.

6

DEEP PLACE

From concepts to praxis

Introduction: another way of designing is possible

For the past decade, a growing number of scholars have been advocating new place-based approaches (see, for example, Marsden, 2018; Horlings et al., 2019; Quinn and Hales, 2021). This is opening up a wider and more active understanding of the potentialities of place-based thinking and ontologies. Escobar (2020, p. 156), for example, suggests:

> Let us practice, for the moment what might be called the architecture of interrelatedness and reconnection, carrying out (perhaps collectively) a great exercise in re-imagining the region – from meticulous new inventories of living forms-flora, fauna, minerals, landscapes – toward a new production of space. Let us situate ourselves in some high place somewhere in the valley (as the legendary Scottish urban planner Patrick Geddes used to do) and ask ourselves: 'what layouts of vegetation, territories, human groups, jobs vocations and professions would produce pleasant configurations of spaces and places, territories and cities? What healthy, comfortable and playful meshworks of infrastructures, cities, humans, and non-humans can we collectively construct? Finally, how can we get this minga (collective labour) going on the ground?'

Escobar introduces a more autonomous social and spatial (pluriversal) theory of development here, which is built upon 'post-western' and post-modernist development principles and new epistemologies. These include non-dualistic thinking and pluriversal development. He proposes innovative re-readings of heteropatriarchy, capitalism, racism, and modernity, promoting dialogues and practices for reconstituting worlds in de-colonial, post-developmentalist, and pluriversal

DOI: 10.4324/9781003221555-7

ways. In this way too, he opens intellectual space for us here in rethinking what types of grounded and more equitable development are needed in different places, so that they could potentially lead to more sustainable and grounded strategies. In Part Two, we will empirically develop these arguments in exploring a rich tapestry of 'Deep Place' studies.

There is a growing institutional and conceptual gap between radical theories of planning and development, and the actual processes of spatial planning practices. The latter, especially in the Global North, are increasingly dominated by a weak form of ecological modernisation, mixed with what seems like an inevitable neoliberal and 'development-led' approach to planning. Whilst this is often advocated as 'bottom-up', inclusive and part of 'sustainable place-making' by the main professional bodies, it only tends to re-enforce standard and exclusionary concerns based upon a local government's financial logic. Such logic is, in turn, based upon 'deals' with land and housing developers to meet the crisis in local government finance, which has been created by reductions in central government funding for public goods and services. Deeply embedded in this policy context is the logic of agglomeration – concentrated development in larger towns and cities (see Chapter 4). London, for instance, with its massive speculative development system is the epitome of this highly selective process, and the horrifying tragedy of Grenfell Tower is an expression of many of its negative outcomes.[1]

As Part One has shown, the realities of current spatial planning processes and outcomes, whilst espousing only a partial incorporation of real sustainable place-making, have tended to be major parts of the problems, rather than solutions, to the combined ecological and economic crisis. In this chapter, we offer a critique of these processes and problems, upon which, a new (or at least revised) and deeper ontological basis for place can be constructed.

Political cultures and place

As we see from the discussion in Chapter 5 (where food is used both as a metaphor and a multi-dimensional materiality for understanding the interactions between place, nature, and cultural practices), an important dimension of the cultural lens that is often ignored in environmental debates concerns the significance of political cultures over extended periods of time, space and place. Swyngedouw (2014) (among others, see: Morangus-Faus and Marsden, 2017), argued that much environmental and ecological debate occurs in a sort of de-politicised context. This has given rise to calls to reinvigorate and expand the role of political ecology perspectives, although, again, this often avoids assumed norms and practices associated with political and policy decision-making systems (for example, planning). The belief that this is de-politicised, of course, is untrue. Instead, neoliberalism has been adeptly penetrating government and state institutions and processes and has appeared to turn politics into a series of technical issues. In fact, politics is now hidden under a veil of neoliberal rationalism that renders many critical policy issues as 'technical'. The environmental field has been particularly affected by this process of de-politicisation. It is, therefore, necessary to understand the nature of the shifting, but still dominant political cultures.

What do we mean by political culture? Possibly the most significant exponent of the importance of political culture was Gramsci, who argued, that at heart we are all politicians and legislators and want to have some level of power (Denning, 2021). The power of the intellectual, the subaltern, is to understand that the art of politics is to stem and manage the contradictions between capitalism, democracy, and social change. This meant that Gramsci provided a far more sociological and political contribution than either Marx or Engels, to understanding the historical transformations that have taken place under capitalism. Different political traditions emerge and evolve within and between political parties, which are often attracted to particular intellectuals and theories of political action. These consequently connect 'downwards' to local and place-based actions and processes. These are often taken for granted and are part of a collective subconscious. In the late 20th century, neoliberalism, for instance, which was based upon the theories of intellectuals like Hayek and Milton Freidman, was increasingly promoted as a legitimate set of theories that were subsequently operationalised into political strategies. The further embeddedness of neoliberalism consequently refined relations between labour and capital, labour and nature, and, as we shall see, place and nature.

These ideas are absorbed, at times, in the relations and contradictions between social liberalism, economic liberalism, and conservatism, which compete regarding their political cultures, assumptions, and practices. Social liberalism, for example, emerged in the late 19th century with intellectuals such as T H Green, D G Ritchie, and L T Hobhouse. This bore the origins of Fabianism and eventually Keynesianism, whereby social democratic solutions were sought to significantly reframe and intervene in the free market tenets of economic liberalism. Successive Conservative and Labour UK governments have since combined elements of the contradictions inherent in balancing these social and economic liberalist ideas (most notably the Blair and Brown governments in the early 2000s). Currently, the UK Conservative party, in theory at least, is placing greater emphasis on economic liberalism as a global and national strategy, and on individualism over collectivism (see Wayne, 2018, 2021). Thus, for Gramsci, the hegemonic superstructure must be managed within the context of conflicting and contradictory organic political cultures and, at the very least, must reach down and out to subalterns for its continued sustenance. This creates problems for the left, as, since the 1980s, economic liberalism continues to dominate global political cultures. In addition, various aspects of social liberalism, such as redistribution and decentralised, democratic decision-making have been strangulated into metamorphic economically liberalist and conservative forms.

In recent UK regional and urban policy, we see this played out in the context of 'City Deals' (see Jones, 2019), 'levelling up' rhetoric, and various aspects of regeneration and infrastructure planning and expenditure. It can also be witnessed in the longer-term pattern of UK reductions in local government financing and consequent cutbacks. This pervasive brand of neoliberal economic policymaking nevertheless relies on large-scale public financing, albeit managed by financial elites. Decision-making, planning, and project management processes create power asymmetries among private and public actors and institutions. There is a cultural assumption of legitimacy in these models of funding and investment decisions made by an economic elite, who enjoy close interactions with government officials, elected representative, and large firms. This is particularly notable in the current triumvirate of the oligopolistic construction, finance, and energy sectors. As Gramsci realised for subalterns, and which the radical left and social liberal environmental movement is now discovering, it is very difficult to disrupt hegemonic structures and political cultures.

So, the reader may ask, what does this discussion of political cultures mean for the place-based sustainability agenda explored in this book? Although often ignored or put to one side by environmental interests, political culture is critical for sustainability discourse. Let us, for example, consider what might be termed an active process of neoliberal 'economic strangulation'. The post-Brexit UK government is developing regional, environmental, and agricultural policies based on assumptions of market-based payments and offsets in the delivery of environmental goods and services. This is especially the case in England, as it remains to be seen whether, and how, the devolved UK nations develop a more socially liberal set of policies. Consequently, economic liberalism continues to penetrate deeper and deeper into the metabolism of environmental restoration. This, we would suggest, has been re-enforced by the Glasgow COP26 outcomes. This is, of course, a contradiction in economic liberalism, as ecological restoration relies upon collective (socially liberal) policies (see Chapter 10). In the absence of dissenting voices in mainstream politics, in the UK parliament at least, it is left to 'organic intellectuals' and campaigners (sometimes termed 'public intellectuals') – such as David Attenborough, Bill McKibben, and Greta Thunberg – to state the case: 'shame on you' or 'do something fast'! Collective and grassroots democratic actions are, of course, anathema to economic liberals (see Purdy, 2015. Meanwhile, social liberals making calls for collective action, real sustainable and democratised state planning, and distributed and place-based solutions and economies, are culturally sidelined by the hegemonic economic liberalist regime.

Thus, as Gramsci informs us, political cultures – resting on often assumed norms of intervention and behaviour – embed themselves and in many ways, or as Marx and Gramsci each suggest 'concretise' themselves. That is, such cultures become the obvious and institutionalised way of doing things and demarcate the radical from the normal and the rational, as irrational. Such processes surround current debates in Wales on the direction of economic and spatial policy (see Lang and Marsden, 2021), whereby Welsh governance currently appears to 'ride two horses' – an economic liberalist growth-focused paradigm and more socially liberalist foundational economy experiments. Each perspective has its own organic intellectuals and subalterns, and each is engaged in a debate over the degree to which socially collective and inclusive actions are politically acceptable.

Ecological and environmental intellectuals, and their widespread global and local networks, have an important role in progressing and establishing alternative, socially liberal, sustainable cultures. Their role in political culture is too often sidelined on the basis that it is the role of such thinkers to 'provide the evidence' and engage in a rational critique and to present this, all nicely wrapped up, for politicians and officials. Moreover, even many people within the environmental movement have come to believe that after over 30 years, the neoliberal political culture is the 'only game in town'. There is, however, a new culture emerging that rejects this linear, economically liberalist cultural system, which is based on methodologies of co-production. The question therefore becomes how can this alternative be assembled and moved forward?

Distributed systems: incorporating circular social and ecological economies

The chapters in Part One have laid the conceptual foundations for a more (as we indicate in Chapter 3) conjoined approach in the treatment of the twin ecological and social inequality crises the world currently faces. As we discussed in these earlier chapters, failures of economic thought and practice have had profound social, environmental, and cultural impacts. Nevertheless, given the will, all may be redesigned.

We argue that a focus on place, as an active and potentially re-empowering concept, can help assemble these more abstract and often separate spheres into powerful perspectives for positive change. Place is, therefore, a conceptual, methodological, and heuristic device. It is not merely a simplified tabular rasa upon which these economic, social, cultural, and environmental interactions are played out. Nor is it simply a spatial consequence of these processes. It is, in various meaningful ways, a highly interactive, contingent, and powerful agent. Consequently, in the second section of this book, we discuss, in detail, a series of long-standing 'deep-place studies', which we have conducted over the past decade to demonstrate how the particular place-based histories and biographies of places help to condition responses to them, as well as to explore the possibilities for new, more sustainable and equitable pathways and trajectories.

A multi-dimensional and multi-species 'more than human' approach to place, requires us to recognise that potential solutions and pathways will be reliant upon the biophysical and socio-natural contexts of particular places. Thus, the geographic features of place must come to the fore, whilst nevertheless recognising the clear significance of more generic and structural forces at play. As Latour (2017) and many others have recognised, this is a place-based world and the forces of globalisation – such as climate change, economic trade and knowledge flows, global geopolitics, pandemics, and various technological innovations – must, therefore, engage in an active process of re-territorialisation. As Latour recognises, the global interrelatedness of territories that re-localisation encourages establishes a new global ethics of care. Moreover, in a world that aspires, however slowly, to decarbonise its food, energy, transport, housing, and major infrastructural systems, we must increasingly recognise that places offer highly differentiated constellations of resources (physical, ecological, and human) that may contribute to these wider goals.

It is not necessary nor desirable, therefore, when focussing on place, to fall into what has been termed 'the local trap', or to become 'localistic' as a defence mechanism against the 'other' (i.e., migrants, external developers, etc.). The local trap denotes simplistic assumptions about only giving importance to sustainability gains through local actions (see Born and Purcell, 2006). Meanwhile, 'defensive localism' (Winter, 2003) suggests the sorts of 'NIMBY' responses witnessed in and around many middle- and upper-class residential communities, in what Murdoch and Marsden (1995) called the local creation of 'class-space', in such areas as South East England's ever-expanding commuter belt. Given these important caveats and provisos, we demonstrate below through empirical and co-produced research with (not on) selected communities, that re-localisation and territorialisation are central features of progressing the principles of 'green new deals' and, more generally, ushering in a new ecological welfarist approach to sustainable well-being (see Chapters 2 and 3).

As we demonstrated in Chapters 4 and 5, this cannot avoid tackling the often-overlooked question of local power structures and property rights configurations, not least because they condition local politics and governance. Over the past decade, many place-based empirical researchers have discovered that community decision-making can lie at the heart of sustainable place-making and community planning (see, for example, Horlings et al., 2019; the 2019 SUSPLACE Programme; and EU Rural Alliances Programme). These initiatives need innovative leadership, existing power elites, and bureaucrats to welcome wider active community participation, and the willingness of 'anchor institutions' to act local in their dealings with communities. This process also presents another local trap – 'local greenwashing'. We recognise the dangers here, not least with some communities that have been the sites for time-limited funding and regeneration projects, which have eventually ended along with corresponding government programmes (see Sjobolm et al., 2012). Nevertheless, these pitfalls can be overcome by developing more supportive governance and political processes that stand the test of time and short-term financial provisioning.

Overall, the profound feature that these transitions encompass is a more collective governance belief in distributed economies, which encourage non-resource extractivism, are socially just, and built upon a cooperative and associational model of interaction between community and policy. What has often been historically castigated as anarchist in character (à la Kropotkin), represent principles of social liberalism that may enhance freedoms through association and effective community-based planning. It tends to be assumed in local and regional economic planning and democracy, as we discussed in Chapter 4, that solutions to the environmental crisis can be conveniently enacted by 'letting the market run the economy'. A more socially and spatially distributed approach, however, must address the undistributed and asymmetrical power and decision-making processes at work, as, for instance, in current UK energy utilities.

Nevertheless, it is possible to witness more distributed decision-making and community planning, not least in the community shares frameworks developing around renewable energy, retailing and a whole host of new small-town services in many of our towns and villages (see Prosser, 2021). As well as, for example, new access being made to formerly restricted land rights by public and community authorities (such as Project Skyline in Wales). Moreover, the Brecon Beacons National Park Authority has now revised its place-based management plan to adopt the Doughnut model of the circular economy and is based on citizen jury community planning initiatives (BBNPA, 2021). Such local innovations need greater support and fostering from national state bodies and agencies for them to realise their true potential. There is always a danger, of course, that such local innovations are either perceived as suspicious or subversive, or as welcome civic engagements that let the national state 'off the hook'. For us, this is clearly the case with the proliferation of food banks in the UK and elsewhere over the period of continued austerity. Many now see these as primary functions of the voluntary civil sector, rather than state-based welfare. Such assumptions feed into a 'self-help' expectation of communities, especially under the severe conditions of state fiscal crisis.

Exploring the anatomy of real sustainable transitions...and avoiding the traps

Part Two documents ways of exploring the 'Deep Place' approach by examining a series of selected cases. It seeks to avoid falling into the aforementioned 'traps' associated with re-localisation and re-territorialisation and raises the question of how places may assemble the means to create real sustainable transitions. There is a well-established literature on sustainable transitions, and we take much of this as a read. Although an uncomfortable stance for many normative and prescriptive scholars in the field of sustainable transitions, our approach is open-ended and non-deterministic. Our focus on place is non-deterministic as, for us, such approaches need to be contingent upon community-based reassessments of place resources in a diverse sense. As such, we must recognise that real sustainable transitions are far more determined by local community and infrastructure factors – the potential strengths of basic or foundational economies, social capital assets, current and future politics, abilities to engage with a range of external markets, and state and political relations – than is commonly assumed. This requires a methodological and ontological approach that is open-ended, unable to predict when and if particular outcomes may occur. This is because such an approach is dependent on the potentialities for change and transition that are created by a particular place's position in relation to the concepts that we outlined in Part One. It is also because the ability of places to adjust to prevailing socio-economic and conjoined ecologies is unequal.

Exploring real sustainable place-making: ontologies of Deep Places

Whilst much of the sustainable place-making literature quite understandably focuses on the present and the future for communities in place (see Franklin, 2018; Marsden, 2018), our foregoing analysis also points to the relevance of understanding and appreciating the cultural, physical, economic, and social politics and legacies that previous rounds of economic activity and regimes of power still play in shaping future strategies and mindsets. In both the Welsh and Australian case studies discussed in Part Two, the legacy of extractive mining casts a long and active shadow over current and future thinking and actions regarding sustainable place-making (see Beynon and Hudson [2021] for a wider UK account of coal extraction cultures). In our rural case study of Llandovery, we see past generations of extractive farming playing a similar role in conditioning and, in some cases, constraining, futures thinking, and actions. Yet, the reinterpretation of local and place-based histories is also becoming a major driver for progressing more sustainable socio-ecological systems and bio-cultural diversity. This is witnessed by Shonil Bhagwat (2018) (see also Doffana, 2018), who demonstrates how, in the context of the Anthropocene[2], indigenous land rights and cultures, ancient traditional practices, and sacred burial sites can become active places for restoring and regenerating places.

As part of Deep-Place thinking there must be a critical appreciation of historical and cultural pasts and layers (acting as palimpsests), many of which may have been actively hidden and distorted during the 20th-century (carbonised and extractive) modernisation, but which now are reappearing. This is now particularly prevalent in scholarly work on the histories of slavery and indigenous societies and is creating a new critical vibrancy across the historical, archaeological, and contemporary humanities and social sciences (see, for example, two notable recent examples of the genre: Holton (2021) and Graeber and Wengrow (2021)). Perhaps, in more contemporary sense, these recasting palimpsests are adjusting themselves in the largely now multi-disciplinary fields of development sociology, anthropology, and geography (as we have noted in Part One). These trends and approaches are highly relevant when we consider the ontologies of Deep Place, and they raise important questions about: what is 'development' and, more uncomfortably, what is desired development? Deep Place approaches must confront these questions head-on.

In our critical treatments of economic, social, environmental, and cultural spheres in Part One, we sought to demonstrate the relative shallowness of the treatment of place. For instance, much economic and environmental thinking has assumed that place, or perhaps more appropriately space, is a largely two-dimensional plane upon which the economic and social division of labour, economic comparative advantage, and the forces of globalisation and agglomeration can be played out. The notion of competitiveness, productivity, and growth are also rooted in these ontologically two-dimensional assumptions. Not least, because the raft of currently dominant economic modelling cannot process the lumpy and multi-dimensional characteristics of nature and place.

Such two-dimensional ontological assumptions are not limited to the realms of economics or the hegemonic political cultures of economic liberalism or conservatism. Unfortunately, the same reductionist and in-built avoidance can be identified (though not universally) in the fields of planning and environmental thinking (as discussed in Chapters 4 and 5). Thus, Deep Place seeks to confront the isotropic and two-dimensional reductionism evidenced in much environmental land use modelling, town and country planning, and resource planning. In the planning profession, this is also a major problem. Major UK planning schools, for instance, have been educating a generation of planners in the field of sustainability, yet once graduating into the professional world of planning and development, the process of institutionalisation tends to dilute and constrain holistic thinking into narrower 'statutory' boxes and silos, not least because of the financial and institutional constraints placed on local planning authorities and the political-cultural assumption that planning should be development-led. Nevertheless, such restricted and closed ontologies are neither acceptable nor appropriate.

Current environmental thinking also tends to elide ontological compatibilities between economic markets as abstract and spatially shallow, whilst planning and development thinking assume that communities must adapt to these assumptions and abstractions. In addition, these convenient ontological compatibilities between reductionist social and environmental science are compounded by a denial or, at best, restricted understanding of deep histories, places, and practices. The fields of anthropology (see Strathern, 2020) and development studies have recognised these dangers and pitfalls for some significant time. What is surprising, therefore, is how the more recent burgeoning fields of environmental, policy, and planning studies, have tended to by-pass or ignore these ontological debates. As such, even honest intellectual avoidance to engage can lead to the implicit assumptions of shallow place thinking. Three examples of recent work help demonstrate the inadequacy and danger of shallow place assumptions.

First, the significant expansion of work over the past 20 years regarding alternative food movements (see Sonnino et al., 2016) and their interconnections with broader political ecology debates (see Morangus-Faus and Marsden, 2017). Sonnino et al. (2016), for instance, argued that in relation to new emerging place-based food geographies:

> The last two hundred years of industrialization and urbanization have promoted, at least in advanced economies, the active and artificial 'flattening' of food geographies, such that, for a long period, we had the 'luxury' of hiding or disguising the significant externalities and inherent diversities associated with the industrial food regime (Marsden and Morley 2014). Resource depletion, climate change and the proliferation of a range of interrelated food insecurities in both industrialized and developing countries are forcing us to re-interrogate this restricted food geography – just at a time when more

segments of society are also growing conscious of its distortions and vulnerabilities. *Globally as well as locally, it is increasingly recognized that we can no longer afford a modernisation project based upon a geographically 'flattened' intensive food system.* Recent reactions to the food security crises have tended to be too fragmented, relying upon (at best) restricted and aggregated geographical conceptions. We now need to recalibrate or even re-create the relationships between the natural and the metabolic with regard to food.

[our emphasis] (p. 30)

Moragues and Marsden (2017) argued in relation to developing the political ecology agenda:

Since 2007, and with the benefit of this scholarly hindsight, we can thus see the need for a re-politicization and socialization of agri-food research in ways *which also opens up the politics and spaces of the possible for a wider range of key actors in the food system.* We have focused in the last part of the paper on two examples: urban food and food sovereignty movements. These dynamic spaces are key to contrast and contest the overwhelming dominant framing of the crisis adopted by the broader natural and economic sciences in the form of a renewed and pervasive neo-productivist reductionism, whereby food security concerns become too narrowly framed around increasing the quantum of production whilst managing sustainability concerns through high rationalist means. At the same time, this dominant framing serves to reduce and marginalize the social, spatial and political basis of food production, processing and consumption to questions of public and consumer acceptability to the onset of more novel scientific techniques in 'solving' the food security and sustainability problems.

[our emphasis] (p. 38)

This process extends well beyond food geographies and relates, more broadly, to a rejection of shallow place perspectives, based upon more empowering place-based approaches.

A significant second area requiring a deeper place perspective is the critique of transition studies, and, in particular, multi-level transition management approaches (see Geels and Schot, 2007; Spaargaren et al., 2012; Lamine et al., 2021). Here transitions have tended to downplay both the spatially embedded nature of transition struggles and pathways, as well as the role of asymmetrical power relations and biophysical relations. As such, these transition approaches have assumed a shallow activity space in which such transitions occur and have overemphasised the linear role of technological innovations in bringing them about. As Rossi et al. (2019) attempted to re-dress these shortcomings, they argued:

Among the existing theoretical-analytical approaches, transition studies,[3] in their diverse perspectives and progressive enrichment through integration with other disciplines, have devoted growing attention to power-related issues to account for the complexity of change processes (Avelino and Wittmayer, 2016). Thanks to their comprehensive approach, they have been intensively adopted to investigate grassroots innovations developing at the micro scale, their interplay with consolidated frames and the associated transformative potential (Seyfang and Smith, 2007; Elzen et al., 2011; Seyfang and Haxeltine, 2012).

(p. 2)

Starting from the Multi-Level Perspective (MLP) (Geels, 2005; Geels and Schot, 2007), based as it is on the dynamics among three levels of structuration (the exogenous pressures of landscape, the dominant institutions and practices of regimes, and the innovative approaches of niches), power relation issues are inherent to transition theories. This heuristic has evolved over time (Geels, 2010 and 2011; Smith and Raven, 2012; Geels, 2014), but continued to be built on the existence of consolidated structures and forms of power, informing socio-technical and governance systems, and spaces of innovation that seek alternatives and aim at transformation. The need to address the role of power more explicitly, has led to further refinements in this framework, leading to a better understanding of agents' power profiles, agency distribution, dynamics between different forms of powers, power shifts among actors, political ontology, and the territorial spatiality of transition politics (Avelino, 2011; Elzen et al., 2011, 2012; Hoffman, 2013; Geels, 2014; Avelino and Wittmayer, 2016; Castàn Broto, 2016; Avelino et al., 2016).

(Rossi et al., 2018, p. 3)

This again reflects the ontological need for place-deepening strategies in transition studies. In each of these critical fields, we have witnessed the realisation that incorporating nature and biophysical aspects of place, not least or only through food systems, requires a radical ontological shift that incorporates re-localisation and Deep Place thinking, without falling into the aggregated and abstractionist traps that orthodox scientific thinking promotes. Thus, Deep Place approaches, as discussed in Part Two avoid these traps and blind spots by focussing on the empirically local in a deep sense, as well as relating this to wider networks of action, practice, and powers of abstraction and aggregation.

A third approach has been developed by the SUSPLACE team, a multidisciplinary EU research programme directed by Ina Horlings and Dirk Roep (2015) (see Horlings, 2018; Horlings et al., 2019). They develop a heuristic tool around embedded place-shaping practices, which involves three key elements that support more deeper place thinking. These are:

- *Re-grounding of practices in place* specific assets and resources to make them more sustainable. Here practices of sustainable place-shaping are influenced by wider communities, cultural notions, values, natural assets, technology, and historical patterns (following Lefebvre [1991], ideas of space as perceived, conceived, and lived).

- *Repositioning towards markets,* referring to the conditions that can enhance the quality of life in places by developing the social economy, social services, alternative products, and markets based on place-based assets. A key question becomes whether these practices have the capacity to be scaled up or out and under what conditions. This echoes the empirical work of Schneider et al. (2016) in Brazil, on the development of alternative food markets as 'nested-markets' (see Marsden et al., 2020), or spatially embedded new markets.
- *Re-appreciation and re-valorisation of the respective places,* showing how actors reflect, enact, and re-negotiate the conditions of engagement with external and global processes to create more autonomy and self-efficacy in place-based development.

As we show in Part Two, such processes can become embedded into a Deep Place approach.

Deep Place thinking conjoins empirical research of micro-relations and practices, with wider knowledge, power structures, and relations of which they are all part (as we identified in Part One). This book is, therefore, an endeavour to build a more ontologically sound approach to sustainable place-making, which attempts to avoid the implicit and sometimes explicit assumptions that take places and their deep histories and cultures for granted. Such assumptions flatten the rich multidimensionality and practices of people, natures, and places. If we are to effectively make the necessary sustainability transition, such open-ended and emancipatory place-based approaches will be essential. This is not a process of inserting one set of reductions (two-dimensional place thinking) with another (place-based approaches). Rather, it is about opening up the ontological and epistemological space for creating foundations of sustainable transitions.

Many scholars will be uneasy about the normative inflections behind these arguments, maybe arguing that it is the scientist's role to describe the world and for others to change it. Yet, as we have seen throughout this book so far and, not least, in this chapter, the role of intellectual thinking has been critical in shaping the problems we face and the solutions proposed. Thus, we should not, as Gramsci and more recently Wright (2010) suggested, draw too many distinctions between theory and practice, but rather examine how they should be critically aligned. Our approaches and arguments in Part One, argue that at the current juncture this is very much a false and dangerous distinction, in that it avoids the point that our current ecological and economic crisis has been founded upon a set of dominant hegemonic theoretical notions, which are now corrupted both intellectually and empirically. We need to construct, as Rousseau argued in the Social Contract (and as Escobar more recently advocates), our own and new experiments, hypothetical and conditional reasonings for living sustainably.

Conclusions: pathways for places

In Part Two, we explore one such 'deep-place' corpus of work. The next chapter (Chapter 7) will outline in more detail the epistemological tools that were used in developing this approach. At its heart, it is an approach that seeks to avoid the traps and blind spots outlined in this chapter, and to fill in some of the onto-logical gaps and opportunities also identified here. It attempts a methodological reconstruction based upon both 'spaces of flows' and 'spaces of and for places'. It is integrative and non-reductionist, in that it applies detailed and forensic diagnos-tics and inclusive solutions to re-creating sustainable community pathways and trajectories. It also raises at least as many questions about the obstacles that exist, as it does to proposing options that may overcome them.

It re-assembles and grounds many of the central concepts discussed in Part One and applies them to practice, including exclusion and poverty, transitions, resilience, basic or foundational economies, and the application of ideas around 'total place'. Hence, it is not sectoral in its approach, linking the main areas of shelter (housing), diet (food), mobility (transport), landscape, nature, work, and leisure (amenity) with the economy. It starts from the premise of wakening us up to the collective and modernist amnesia, which assumes the economy and nature can be developed in historically different pathways. Ultimately, it suggests that there is hope for places and their peoples, putting both centre stage in creating sustainable livelihoods and communities.

In attempting to bridge both the theoretical and conceptual (Part One) and build a new praxis (Part Two), the approach is challenging and ambitious. It is, however, necessary if we are to engage and empower communities and collectives to cope with the Anthropocene, and to redevelop political cultures and economies of care and mutual, collective sustainability. The political and policy world emerging from the UN Sustainable Development Goals and the series of COP26 exercises is dom-inated by the quest to secure and appeal for collective actions. These arguments are weak, however, if they are only top-down, and, as we show, partial, sectoral, and shallow spatial policies are advocated. The action-based research that is out-lined in the succeeding chapters is much needed to reverse the ends of this current Anthropocenic 'telescope'. As we discussed in Chapter 10, there is an urgent need to actively consider the rooted micro, to achieve and repopulate the macro; and to replace excessive and narrow economic individualism, with a rebalanced and place-based set of ecological and economic collective ethics based upon natural and social empowerments. These are some of the lessons the authors wish to share from their conceptual and empirical experiences in diverse places over the past decade.

Notes

1 The fire at the high-rise residential block killed over 70 people in 2017, a result of cheap and inflammable external cladding, and poor fire safety measures. It is in one of

London's and UK's richest boroughs, but experienced severe overcrowding, multiple occupation, and deprivation.

2 The Anthropocene is the term used in wider scientific debates surrounding the new geological time zone following the Holocene. Emerging initially from the geological community it proposes that humans now have an explicit effect upon the more recent geological time and sedimentation series, and this is and will in the future be recorded as a phase when 'man-made' carbonised and industrial effects will be demonstrated in geological history.

3 Transition studies are aimed at understanding broad and deep socio-technical transformations, including, but not limited to, agro-food systems

References

Avelino F (2011) *Power in transition: Empowering discourses on sustainability transitions* (PhD-thesis). Rotterdam: Erasmus University.

Avelino F, Grin J, Pel B and Jhagroe S (2016) 'The politics of sustainable transitions', *Journal of Environmental Policy and Planning*, 18(5): 557–567.

Avelino F and Wittmayer J M (2016) 'Shifting power relations in sustainability transitions: A multi-actor perspective', *Journal of Environmental Policy and Planning*, 18(5): 628–649.

Beynon H and Hudson R (2021) *In the Shadow of the Mine*. London: Verso.

Bhagwat S (2018) 'Introduction: What nature and which society? The complexities of nature-society relationships in the Anthropocene', pp. 935–938 in: T K Marsden (ed.) *Part Ten in The Sage Handbook of Nature*. London: Sage.

Bocci R (2014) 'Seeds between freedom and rights', *Scienze Territorio*, 2: 115–121.

Bocci R, Pearce P and Chable V (2014) 'Policy recommendations for legal aspects of seed certification and protection of Plant Breeders' Rights and Farmers' Rights': Accessed January 2022: www.solibam.eu/modules/wfdownloads/visit.php?cid=14&lid=54SOLIBAM

Born B and Purcell M (2006) 'Avoiding the local trap: scale and food systems in planning research', *Journal of Planning Education and Research*, December 2006, 3: 1–20.

Brecon Beacons National Park Authority (BBNPA) (2021) *Future Beacons Draft Consultation Management Plan 2022–2027*. Brecon: BBNPA.

Bui S (2015) *For a territorial approach of ecological transitions. Analysis of an on-going transition towards agroecology in Biovallée (1970–2015)* (PhD–thesis). AgroParisTech.

Bui S, Cardona A, Lamine C and Cerf M (2016) 'Sustainability transitions: insights on processes of niche-regime interaction and regime reconfiguration in agro-food systems', *Journal of Rural Studies*, 48: 92–103.

Castan Broto V (2016) 'Innovation territories and energy transitions: energy, water and modernity in Spain, 1939–1975', *Journal of Environmental Policy and Planning*, 18(5): 712–729.

Denning M (2021) 'Everyone is a legislator', *New Left Review*, 129: 29–47.

Doffana Z (2018) 'Chapter 49: The role of sacred natural sites in conflict resolution: lessons from the Wonsho sacred forests of Sidama, Ethiopia', pp. 938–968 in: T K Marsden (ed.) *The Sage Handbook of Nature*. London: Sage.

Elzen B, Geels F W, Leeuwis C and van Mierlo B (2011) 'Normative contestation in transitions "in the making": Animal welfare concerns and system innovation in pig husbandry', *Research Policy*, 40: 263–275.

Elzen B, van Mierlo B and Leeuwis C (2012) 'Anchoring of innovations: assessing Dutch efforts to harvest energy from glasshouses', *Environmental Innovation and Societal Transitions*, 5: 1–18.

Escobar A (2020) *Pluriversal Politics: The Real and the Possible.* Durham, NC: Duke University Press.

Franklin A (2018) 'Introduction to part three: spacing natures: resourceful and resilient community environmental practice', pp. 267–284 in: T K Marsden (ed.) *The Sage Handbook of Nature.* London: Sage.

Geels F W (2005) *Technological Transitions and System Innovations: A Co-Evolutionary and Socio-Technical Analysis.* Cheltenham: Edward Elgar Publishing

Geels F W and Schot J (2007) 'Typology of sociotechnical transition pathways', *Research Policy*, 36(3): 399–417.

Geels F W (2010) 'Ontologies, socio-technical transitions (to sustainability), and the multi-level perspective', *Research Policy*, 39(4): 495–510.

Geels F W (2011) 'The multi-level perspective on sustainability transitions: responses to seven criticisms', *Environmental Innovation and Societal Transitions*, 1(1): 24–40.

Geels F W (2014) 'Regime resistance against low-carbon transitions: introducing politics and power into the multi-level perspective', *Theory, Culture & Society*, 31(5): 21–40.

Gibson-Graham J K (1996) *The End of Capitalism (As We Knew It).* Minneapolis: University of Minnesota Press.

Gibson-Graham J K (2006) *A Postcapitalist Politics.* Minneapolis: University of Minnesota Press.

Graeber D and Wengrow D (2021) *The Dawn of Everything: A New History of Humanity.* New York: Farrar, Straus and Giroux.

Gramsci A (1998) *Selections from the Prison Notebooks.* London: Quinton House and Geoffey Nowell Smit.

Hoffman J (2013) 'Theorizing power in transition studies: the role of creativity in novel practices in structural change', *Policy Sciences*, 46(3): 257–275.

Holton W (2021) *Liberty is Sweet: The Hidden History of the American Revolution.* New York: Siomon and Shuster.

Horlings L G and Roep D (2015) *Project Proposal Marie Curie ITN.* Wageningen: Wageningen University.

Horlings L G (2018) 'Chapter 16: Politics of connectivity: the relevance of place-based approaches to support sustainable development and the governance of nature and landscape', pp. 304–324 in: T K Marsden (ed.) *The Sage Handbook of Nature.* London: Sage.

Horlings I, Roep D, Mathijs E and Marsden T K (2019) 'Exploring the transformative capacity of place-shaping practices', *Sustainability Science*, 15: 353–362.

Ingram J, Curry N, Kirwan J, Maye D and Kubinakova K (2014) 'Interactions between niche and regime: an analysis of learning and innovation networks for sustainable agriculture across Europe', *Journal of Agricultural Education and Extension*, 21(1): 55–71.

Jones M (2019) *Cities in Crisis: The Politics of Sub-National Economic Development.* Cheltenham: Edward Elgar.

Lamine C, Magda D, Rivera-Ferre M and Marsden T K (eds.) (2021) *Agro-Ecological Transitions, between Determinist and Open-Ended Visions.* Brussels: Peter Lang,

Lang M and Marsden T K (2021) 'Territorialising sustainability: decoupling and the foundational economy in Wales', *Territory, Politics and Governance*. Autumn 2021 (online): 1.14. DOI 10.1080/21622671.2021.1941230.

Latour B (2017) *Facing Gaia: Eight Lectures on the New Climatic Regime.* Cambridge: Polity Press.

Lefebvre H (1991) *The Production of Space.* Oxford: Oxford University Press. (Originally published 1974).

Marsden T K and Morley A (eds.) (2014) *Sustainable Food Systems: Building a New Paradigm.* Abingdon: Earthscan/Routledge.

Marsden T K (ed.) (2018) *The Sage Handbook of Nature*. 3 Volumes. London and New York: Sage.

Marsden T K, Lamine C and Schneider S (2020) *A Research Agenda for Global Rural Development*. Cheltenham: Edward Elgar.

Moragues-Faus A and Marsden T K (2017) 'The political ecology of food; creating spaces of possibility in a new research agenda', *Journal of Rural Studies*, 55: 275–288.

Murdoch J and Marsden T K (1995) *Reconstituting Rurality: Class, Community and Power in the Development Process*. London: University College London Press.

Pimbert M (2006) *Transforming Knowledge and Ways of Knowing for Food Sovereignty*. London: International Institute for Environment and Development.

Prosser J (2021) *Place-based rural development and cooperative structures: An exploration of the Community Shares Model and its effects on rural communities*. Cardiff University PhD Thesis.

Purdy J (2015) *After Nature: A Politics of the Anthropocene*. Cambridge, MA: Harvard University Press.

Quinn M and Hales T (eds.) (2021) *Our Legacy. Sustainable Places Research Institute, Final Report*. Cardiff: Cardiff University.

Renting H, Schermer M and Rossi A (2012) 'Building food democracy: exploring civic food networks and newly emerging forms of food citizenship', *International Journal of Sociology of Agriculture and Food*, 19: 289–307.

Rossi A, Bocci R, Bussi B, De Santis G, Franciolini R and Pozzi C (2018) 'New goals, roles and rules around agrobiodiversity management', *Proceedings of 55° Conference of SIDEA*. Perugia, Italy.

Rossi A, Marsden T K and Bui S (2019) 'Re-defining power relations in agri-food systems', *Journal of Rural Studies*, 8: 147–158.

Rotmans J and Loorbach D (2010) 'Towards a better understanding of transitions and their governance: a systemic and reflexive approach', pp. 105–222 in: J Grin, J Rotmans and J Schot (eds.) *Transitions to Sustainable Development; New Directions in the Study of Long-Term Transformative Change*. New York: Routledge.

Schneider S, Salvate N and Cassol A (2016) 'Nested markets, food networks and new patterns of food networks in Brazil', *Agriculture*, 6(4): 1–19.

Seyfang G and Smith A (2007) 'Grassroots innovations for sustainable development: towards a new research and policy agenda', *Environmental Politics,* 16: 584–603.

Seyfang G and Haxeltine A (2012) 'Growing grassroots innovations: exploring the role of community-based social movements in sustainable energy transitions', *Environment and Planning C*, 30(3): 381–400.

Sjobolm S, Andersson K and Skerratt S (eds.) (2012) *Sustainability and Short-Term Policies: Improving Governance in Spatial Policy Interventions*. Abingdon: Routledge.

Smith A (2007) 'Translating sustainabilities between green niches and socio-technical', *Technology Analysis and Strategic Management*, 19(4): 427–450.

Smith A and Raven R (2012) 'What is protective space? Reconsidering niches in transitions to sustainability', *Research Policy*, 41: 1025–1036.

Sonnino R, Marsden T K and Moragues-Faus A (2016) 'Relationalities and convergences in food security narratives: towards a place-based approach', *Transactions of the Institute of British Geographers*, 41(4): 477–489.

Spaargaren G, Oosterveer P and Loeber A (2012) *Food Practices in Transition: Changing Food Consumption, Retail and Production in the Age of Reflexive Modernity*. Abingdon: Routledge.

Stock P V, Forney J, Emery S B and Wittman H (2014) 'Neoliberal natures on the farm: farmer autonomy and cooperation in comparative perspective', *Journal of Rural Studies*, 36: 411–422.

Strathern M (2020) *Relations: An Anthropological Account.* Durham, NC: Duke University Press.

Swyngedouw E (2014) 'Insurgent architects, radical cities and the promise of the political', pp. 169–188 in: J Wilson and E Swyngedouw (eds.) *The Post-Political and Its Discontents.* Edinburgh: Edinburgh University Press.

Wayne M (2018) *England's Discontents: Political Cultures and National Identities.* London: Pluto Press.

Wayne M (2021) 'Roadmaps after Corbyn: parties, classes, political cultures', *New Left Review,* 131: 37–67.

Winter M (2003) 'Embeddedness, the new food economy and defensive localism', *Journal of Rural Studies,* 19(1): 23–32.

Wright E O (2010) *Emerging Real Utopias.* Chicago, IL: Chicago University Press.

PART TWO

PART TWO

7

UNDERSTANDING PLACES

Introducing Deep Place

Introduction

Deep Place is a methodology that allows a thorough analysis and understanding of a place, upon which planning and policymaking can be formulated. A place-based technique, it is grounded in an empirical concern with how to achieve more economically, socially, environmentally, and culturally sustainable places and communities. It is based on the premise that a properly functioning economy should add to, rather than undermine, the social, environmental, and cultural sustainability of places and communities. In this respect, Deep Place is our attempt to apply the broader, more theoretical discourse contained in Part One in individual communities. Tim Jackson (2019) summarised Manuel Castells' (1996) work *The Rise of the Network Society* in which 'he suggests that those who function within the dominant networks are privileged to be a part of the 'space of flows' (i.e., global) and their situation contrasts with the excluded, who live in the 'space of places' (i.e., local)' (p. 212). Deep Place seeks to redress the balance of power in these relationships by putting place (the local) at the centre of sustainable place-making.

As we outline in this chapter, such an approach ensures that communities can fully engage in pinpointing and planning for their own distinctive future needs, identify locally grounded opportunities to support sustainable local economic activity, and define and safeguard their cultural distinctiveness. In this model, the function of regional and national government becomes a supportive and facilitatory one. This, of course, does not undermine the role of the state (often acting in collaboration with other agencies), as we see it, to ensure fundamental standards and objectives (for example, equality, redistribution, democracy, environmental standards), but that better distribution of power is achieved, and that, consequently, the policy is more responsive, better designed, and sustainable.

DOI: 10.4324/9781003221555-9

In Chapters 8 and 9 that follow, we provide summaries of the practical experience of undertaking Deep Place studies in six communities spread across Wales, Australia, and Vanuatu. In this chapter, we offer a practical commentary on how these studies have been undertaken. This chapter firstly summarises some of the key methodological influences and then outlines the specific research methods that have been applied in each of the case studies.

Methodological influences

Social exclusion

In 2016, the Joseph Rowntree Foundation (JRF) calculated that the total cost of poverty in the UK stood at around £78 billion, a significant proportion of which was accounted for by public service expenditure used to deal with the effects of poverty (£1 in every £5 spent on public services) (Bramley et al., 2016). Historically labour market exclusion was the largest cause of poverty in the UK, but, as the JRF has shown, 'in-work' poverty has overtaken it (JRF, 2013, 2020). Working-age adults living in families where at least one person works now account for 58 percent of all adults living in poverty. Meanwhile, children living in working families now account for 73 percent of all children living in poverty (Francis-Devine, 2021). This, of course, does not overlook the fact that both adults and children living in non-working households are far more likely to experience poverty, but it does illustrate the fact that work is no longer a guarantee to avoid poverty in the UK. This pattern is replicated in many other advanced economies.

As we discussed in detail in Chapter 3, social exclusion analysis is an important component of sustainable place-making strategies. The experience of poverty tends to be higher in post-industrial and inner-city communities. The percentage of people living in relative low income *before* housing costs in the UK is highest in the North East, Yorkshire and Humber, and Wales, whereas *after* housing costs London has the largest percentage of people living in relative low income (Francis-Devine, 2021). Poverty in rural areas tends to be less visible than urban poverty. Partly, this is due to the more dispersed nature of rural settlements, but it is also a result of some divergence between rural deprivation and traditional characteristics associated with urban poverty. Rural poverty tends to be more closely associated with low wages than with unemployment, as is the case in urban areas; low-income households in rural areas are also more likely to consist of two-parent families and contain older people; and the built environment also tends not to conform to the same characteristics found in dense urban areas where poverty predominates (Milbourne, 2011).

Social exclusion helps explain patterns of economic and social disadvantage in both urban and rural contexts by demonstrating why certain communities confront barriers that prevent them from participating in economic, social, and political life. Social exclusion is multidimensional, and very much context-dependent (see Chapter 3). As we have discussed in previous chapters, economic development and economic growth have not been inclusive and they have failed to reduce the deficits in rewarding work. Since the mid-1990s, not only has economic growth failed to address social exclusion, but it may have also been contributing to it. Indeed, as the United Nations (2016) has highlighted 'the risk of holding a poorly paid, precarious or insecure job is higher today than it was in 1995' (p. 3). Moreover, trends towards greater social progress in the immediate aftermath of the Second World War appear to have gone into reverse.

In the UK, the Augar (2019) report on post-18 education found that there had been '...no improvement in social mobility in Britain over half a century...[and] increases in wealth and changes in the overall structure of the job market have had no impact on the relative chances of people born into less advantaged groups' (p. 23). Research for the Welsh further education sector (Lang, 2020) has found that place has a significant impact on overall educational attainment and that this is because spatial economic inequalities tend to be mirrored in a socio-economically polarised schools' system. Attainment rates are heavily skewed towards more affluent locations, and whilst the report found that there had been attainment growth across the whole of the UK over the previous 20 years, such growth had been strongest in regions that already had a more highly qualified workforce.

Transition theory

Much reported during the COP26 United Nations Climate Change Conference in 2021, transition thinking has predominantly been concerned with the supply and use of energy from fossil fuels and consequential carbon emissions that cause climate change. Ultimately, a draft commitment to end the use of coal was watered down to a less conclusive commitment to reduce the use of coal. Whilst much of the COP26 was focused on the use of coal, transition thinking asks broader questions about what societies must do to manage a successful transition in a wider sense. In a process that has been referred to as 'backcasting', transition thinking asks *what kind of future state do we wish to create, and what must we do now to achieve it in a planned way?* Richard Heinberg (2008) suggested that the future state is likely to be one in which, because of the need to cut carbon emissions, rising costs, and the increasing scarcity of fuel, the transportation of goods becomes more problematic. Consequently, he suggested that economic re-localisation would become inevitable and that as we will need to produce more of our basic goods closer to where they are needed, the immediate community should become the source and focus of transition strategies. For us, this has not meant local economies can (because global interconnectedness is far too complex to unpack entirely) or should (as clearly it would be illogical for *all* goods and services to be produced in every community) exist in isolation, but that more essential goods and services should be produced closer to the point they consumed.

Resilience – a state where communities have far greater control of the productive processes that are required to meet their basic needs – is a core concern of the transition movement. Resilience therefore represents the ability of communities to continue to function despite external shocks and longer-term pressures, notwithstanding the need for some redistribution, particularly where local production is negatively impacted because of an internal or external shock. Interest in fuel and food security, and concerns about reliance on extended supply chains, derive from this idea of resilience. The weaknesses of current supply chain arrangements were, of course, exposed to a far greater extent during the COVID-19 crisis than at any other time in the post-war period. Although some have criticised the transition movement for its apparent separation of citizen and state, and for regarding aggregated individual actions as a source of global change (see North and Longhurst, 2013), we believe it is possible to enrich transition thinking by a greater duality of citizen and state. Whereas a successful transition will certainly require the citizen to adopt a less consumer-driven lifestyle, the central and local states will need to occupy an important position to develop low carbon infrastructure and incentivise sustainability. The state also has a critical role to play in ensuring that the required transition is equitable and does not reinforce existing spatial inequalities.

Interest in 'just transitions' has been growing. In 2019 the Scottish Government, for example, has established a Just Transition Commission to develop co-produced plans that aim to create a net-zero economy in Scotland that tackles, rather than adds to inequality and injustice. Its interim report, published in 2020, recognised that whereas the transition debate has been focused on the creation or destruction of jobs, the consideration of equity needs to be much broader. The UN's COP26 Climate Change Conference agreed on the need to support the conditions necessary for a just transition internationally, and the need to diversify to create more sustainable, resilient, and inclusive economies, rather than simply replacing one industry with another. The Conference ultimately failed to fundamentally address the questions of global inequality, however, and did not resolve tensions over who pays for transition in poorer nations, particularly when such transitions are required for global well-being.

Total Place

Total Place was a term applied to an approach to public service reform developed by the UK's Leadership Centre for Local Government, which, between 2008 and 2010, undertook 13 pilot projects across England. The pilot projects attempted to calculate total public spending in each of these areas, sought to change the way public services were provided, and to devolve delivery responsibility to each area. Grint and Holt (2011, p. 87) concluded that the Total Place approach represented a significant shift from previous practice, as its focus was '...the person/citizen/customer and prevention through partnership, rather than the service/silo response that has hitherto prevailed'. Total Place attempted, therefore, to develop an approach to public service reform that was place-based, meeting the needs of particular communities, rather than being driven by siloed and abstract policy implementation. The Total Place experiments acknowledged that no single approach or method could be suitable for all circumstances and that each place must therefore determine the kinds of actions that were necessary to create higher public value in the provision of public services that better meet the needs of recipients, whilst improving the performance and achieving greater accountability of service providers (Grint, 2009).

As social problems have become more complex or 'wicked', the connections between cause and effect are less clear, and therefore abstract and siloed solutions are unlikely to succeed. Wicked problems – for example, obesity, drug abuse, violence, and anti-social behaviour – have complex socio-economic causes that sit across and between different government departments and institutions. At their core, the Total Place experiments were an attempt to address the difficulties experienced by public services in overcoming these complex social problems. By adopting a systems-based approach to consider the connections between the silos and the services, Total Place sought to achieve a richer picture of the causes of complex social problems. A national system of public provision, it was argued, was far too complex to map, but a clearer 'Total Place' picture could emerge at a local level, particularly if local knowledge was embedded into the process by '…"listening" to people, either recipients of services or deliverers, in order to try and understand the issues from multiple perspectives' (Grint and Holt, 2011, p. 93).

Fundamentally, Total Place was informed by several existing theories, including chaos theory, systems theory, change theory, and leadership theory. Based on these, Total Place attempted to develop an approach that could deliver public services to communities in a holistic, integrated, and collaborative way. It was hoped that such an approach would not only change the way that services were delivered but that overall public expenditure would be refocused from symptoms to causes. The aim was to focus on 'real' projects, move beyond traditional partnership approaches, and ask 'exactly what do our customers need and are we delivering it?' (Leadership Centre for Local Government, 2010). Beyond 2010, with a change in UK Government such approaches were not progressed further in any meaningful sense in England. Instead, the focus shifted towards notions of 'big society' and 'localism', which embodied a much-reduced role for public services that appear to be counter to Total Place approaches, and that eventually morphed into growth-focused programmes such as City Deal.

In Wales, following the Well-being of Future Generation Act, elements of the Total Place approach can be found in the creation of Public Service Boards and the necessity for county-wide well-being plans. Although many of the critical elements, such as joint budgeting, have yet to fully emerge, the Public Service Boards do offer a potential forum to develop Total Place-type thinking, particularly regarding the provision of public services. Nonetheless, whereas the English model has moved away from a focus on the public to private sector, the Welsh Public Sector appears to play insufficient regard for the private sector. Public Service Boards are composed of representatives from public and third-sector partners, who are developing and implementing well-being plans with limited reference to private sector partners. In neither case has there been sufficient engagement with communities in place planning.

The Foundational Economy and anchor institutions

Reflecting on how economic policy tends to focus on sectors that have been 'framed' as having potential for growth, development, and export potential, the Centre for Research on Social and Cultural Change (CRESC) coined the term 'Foundational Economy' (Bentham et al., 2013). The problem, for CRESC, was that these chosen sectors – for example, aero-space, pharmaceuticals, and hi-tech industries – have little potential to build stable and sustainable economies and are, in any case, spatially imbalanced. In part, this is because the sectors favoured by contemporary economic policy represent only a small proportion of the overall economy; but also, because governments tend to lack the policy levers necessary to ensure these sectors become primary economic motors. This latter issue is amplified because of competition from other advanced economies pursuing similar strategies. CRESC therefore argued for more balance in economic policy and, consequently, for greater focus on the neglected Foundational Economy, which they defined as '...producing welfare-critical goods and services like housing, education, childcare, healthcare and utility supply' (Froud et al, 2018, p. 19).

The Foundational Economy appears to echo Blumenfeld's (1955) discourse on the 'basic-nonbasic concept', where 'the basic economy' consists of those activities that citizens supply one another. From this perspective, '...urban development policy should therefore focus not only on the inherently volatile segments of the urban economy but also on the economic activities that maintain the daily functioning of the metropolis and its citizens' (De Boeck et al., 2019, p. 75). Significant interest in the basic or Foundational Economy has emerged, particularly in Wales, where the Welsh Government has acknowledged its importance in its most recent economic strategy (Welsh Government, 2017), and has launched a Foundational Economy challenge fund to progress experiments in this part of Wales' economy.

Alongside this interest in the Foundational Economy, there has been a growing recognition of the role played by 'anchor institutions'. This is a separate development in the literature, but, because of obvious common interest in supporting the development of local economies for social good, the boundaries between these two kinds of literature are often confused. Whereas anchor institutions have a potentially significant role to play in foundational economies, Foundational Economy thinking is not entirely coterminous with, nor entirely dependent upon, anchor intuitions. Taylor and Luter (2013) suggested that deindustrialisation, the globalisation of capital, and the decline of town centres have created precarious social and economic contexts, in which certain 'anchor' institutions have become sources of stability. Such institutions were first identified in the USA during the late 1960s, when many cities began to experience urban decay. Initially, interest was largely confined to the anchor role of universities and health care providers, but this has subsequently expanded to other types of organisations. Harkavy and Zuckerman (1999) suggested that towns and cities had a range of 'fixed assets' that could help regenerate them. These assets were the large and spatially immobile institutions that impacted local economies through their procurement practices, investment, real estate development, business incubation, and, particularly in the case of universities, purchases made by staff and students (Hodges and Dubb, 2012). Birley (2017) suggested there is a need for networks of local and regional anchors, involving basic frameworks of behaviours to support local economic development, participation in community wealth building, and local employment.

Discussions around how to better support and develop local foundational economies, as well as enhancing the role of anchor institutions, appear to offer much to a discourse concerned with sustainable place-making. For us, as we have discussed elsewhere in this book, the re-localisation of economic activity is a critical part of Deep Place planning. However, although the local economic activity is more likely to contribute to wholistically sustainable transitions than large, globalised supply chains, localisation is no guarantee of sustainability. Indeed, it has been suggested that Foundational Economy thinking pays insufficient regard to ecological imperatives (Hansen, 2021). For us, therefore, these discourses are necessary elements, but not sufficient in sustainable place-making.

Sustainable place-making and the UN Sustainable Development Goals

Issues of social equity and climate change mitigation, driven by the sense of crisis in the social and ecological domains experienced globally, have occupied a central position within the Deep Place approach. Addressing these issues, in 2015 the United Nations General Council agreed Resolution 70/1, *Transforming our World: The 2030 Agenda for Sustainable Development* (United Nations, 2015). The aims of the resolution are encapsulated in the phrase: 'we are resolved to free the human race from the tyranny of poverty and want and to heal and secure our planet' (p. 1). The resolution announced 17 Sustainable Development Goals (SDGs) and 169 associated targets. Now familiar as the colourful icons of hope for greater social and climate justice, the goals were developed from the social objectives of the UN's *Millennium Development Goals*, to include aspirations of climate change mitigation and planetary environmental recovery. Although adopted by 193 countries, the SDGs have thus far received variable degrees of support and achieved mixed rates of progress. A recent report (Sachs et al., 2021) illustrating progress made by countries towards meeting the SDGs, shows that three Scandinavian nations currently occupy the top positions (Finland, Sweden, and Denmark are ranked first, second, and third, respectively). Meanwhile, those countries where the various Deep Place case studies have so far been undertaken are ranked 17th (UK), 35th (Australia), and 119th (Vanuatu).

The COVID-19 pandemic had a major impact on progress towards the SDGs, with an estimate of between 119 and 124 million people entering extreme

TABLE 7.1 UN Sustainable Development Goals

1	End poverty in all its forms everywhere
2	End hunger, achieve food security and improved nutrition, and promote sustainable agriculture
3	Ensure healthy lives and promote well-being for all at all ages
4	Ensure inclusive and equitable quality education and promote lifelong learning opportunities for all
5	Achieve gender equality and empower all women and girls
6	Ensure availability and sustainable management of water and sanitation for all
7	Ensure access to affordable, reliable, sustainable, and modern energy for all
8	Promote sustained, inclusive, and sustainable economic growth, full and productive employment, and decent work for all
9	Build resilient infrastructure, promote inclusive and sustainable industrialisation, and foster innovation
10	Reduce inequality within and among countries
11	Make cities and human settlements inclusive, safe, resilient, and sustainable
12	Ensure sustainable consumption and production patterns
13	Take urgent action to combat climate change and its impacts
14	Conserve and sustainably use the oceans, seas, and marine resources for sustainable development
15	Protect, restore and promote sustainable use of terrestrial ecosystems, sustainably manage forests, combat desertification, and halt and reverse land degradation and halt biodiversity loss
16	Promote peaceful and inclusive societies for sustainable development
17	Strengthen the means of implementation and revitalise the Global Partnership for Sustainable Development

poverty in 2020, the first increase in 20 years (UN, 2021). Even without the COVID-19 crisis, however, there is a general sense that progress on the SDGs and Agenda 2030 was unlikely to meet the 2030 target. A 2020 United Nations report concluded that:

> ....before the COVID-19 pandemic, progress remained uneven and we were not on track to meet the Goals by 2030. Some gains were visible: the share of children and youth out of school had fallen; the incidence of many communicable diseases was in decline; access to safely managed drinking water had improved; and women's representation in leadership roles was increasing. At the same time, the number of people suffering from food insecurity was on the rise, the natural environment continued to deteriorate at an alarming rate, and dramatic levels of inequality persisted in all regions. Change was still not happening at the speed or scale required.
>
> *(United Nations 2020, Foreword)*

The SDGs are reinforced by the New Urban Agenda (NUA) (United Nations, 2017), agreed upon by 167 nations at the Habitat III conference in Quito,

Ecuador in 2016. Whilst less visible (and visual) than the SDGs, the NUA is seen as an accelerator of the SDGs in cities and urban places. The NUA effectively embodies the aspirations of SDG 11, *Sustainable Cities and Communities*, and tends to be regarded as a mechanism for localising Agenda 2030 and identifying actions at the city level to achieve the SDGs. Although the first Deep Place study was published in 2014, and therefore predated these UN agreements, the method aligns well with both the SDGs and the NUA. As we have indicated, the core objectives of the Deep Place approach are to resolve poverty by developing a climate-sensitive economy, and its concern with social, economic, environmental, and cultural sustainability equips it to support local actions required to meet the SDGs and NUA.

Research methods

It is important to note that the research methods utilised in each of the Deep Place case studies that are summarised in Chapters 8 and 9 have been dynamic. Whereas the methods outlined below have tended to be deployed in each study, particular methods have been utilised to greater or lesser extents, or added as new elements, because of experience gained from undertaking each successive study. Nevertheless, certain fundamental principles have been consistent in each study. The methods used have been both desk-based quantitative, and ethnographic qualitative. The research methods adopted in each study have been inclusive of the wide variety of agencies and organisations involved in the various policy areas impacting each community, and which are therefore critical to achieving a successful transition to a more sustainable and equitable future model for each community. Community participation has also been a consistent and central element of each study.

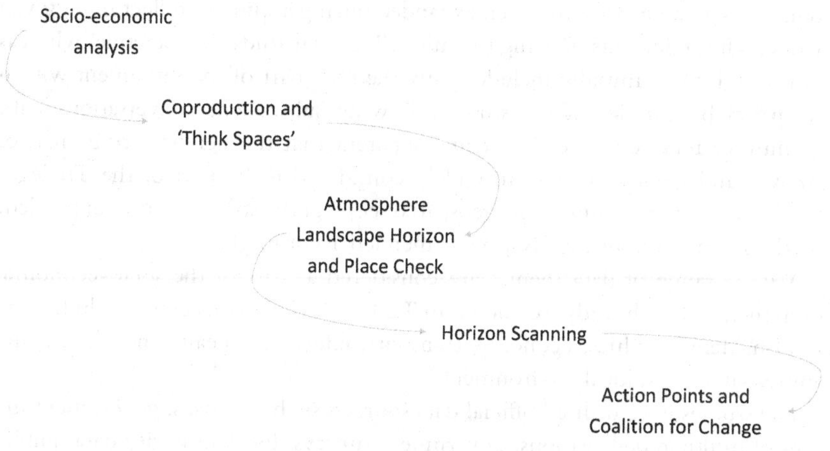

Socio-economic
analysis

Coproduction and
'Think Spaces'

Atmosphere
Landscape Horizon
and Place Check

Horizon Scanning

Action Points and
Coalition for Change

FIGURE 7.1 Research structure utilised during a Deep Place study.

Deep Place is action-based research, which seeks to influence policy and delivery, and each case study has sought to be a catalyst for change in internal and external community perceptions. There have been consistent attempts to involve many of those actors and agencies who would be critical to implementing the transition advocated in each report during the research phase. Whereas some locations have been 'selected' by the research team, others have been requested by one of more agencies practically involved in the well-being of the community. Indeed, prior to the commencement of each study, the research team has sought to engage the support of local authorities and key agencies, and we have found that overall, this has been actively forthcoming. No study has been undertaken where there have been local objections. Only in one instance has the initial consultative process led to a study being undertaken elsewhere, and this was a result of an agency already undertaking its own place-planning process and therefore wishing to avoid local confusion. Regardless of how each study emerged, the research methods utilised have sought to embody the principles of coproduction within the research process, principally through the use of 'Think Spaces' (see below), so that that the community and various public, private, and social partners have a sense of ownership of the reports produced and action points recommended.

Socio-economic analysis

Each Deep Place study commences with an extensive, statistically informed socio-economic analysis of the community. Typically, this begins with identifying the geographical boundaries that will be used for each area and, following this, the relevant geospatial data from which evidence is drawn to undertake the analysis. There is no set spatial or population size limiting the scope of a Deep Place study. Some studies have included local populations of around 18,000, whereas others have had fewer than 2,000 residents. On occasion, this initial-boundary setting process has been expanded during a study to reflect input given by research participants. During the initial Tredegar study, for example, whereas the research team initially included only the core part of the settlement within the spatial boundaries of the study, following subsequent conversations with community representatives it become apparent that the neighbouring areas of Sirhowy and Georgetown were widely considered to be part of the Tredegar settlement and community. The geospatial scope of the study was thus expanded, and the socio-economic analysis was amended accordingly.

A wide range of data themes are considered as part of the socio-economic components of each study. As shown in Table 7.2, issues considered include deprivation, demographics, economy, transport, education, health, housing, community safety, and local environment.

The studies have utilised official data sources, such as census, government indices of multiple deprivations, government surveys, local authority data, public health authority data, and council tax and council tax bands data. Census data

TABLE 7.2 Data themes used as part of a Deep Place study socio-economic assessment

Deprivation	Indices of multiple deprivation, small area income estimates, child poverty rate
Population	Age profile, population density, country of birth, ethnic group, lone parents, teenage pregnancies
Economy	Unemployment, employment by sector, socio-economic classification, business counts
Transport	Method of travel to work, public transport availability
Education	Highest level qualifications, school leavers with skills and qualifications, school absences, participation in post-16 education
Public health	Life limiting conditions, public health, low birth weights, child obesity, life expectancy at birth
Housing	Tenure, accommodation type, property values, council tax bands
Community safety	Crime rates
Environment	Local emissions

has been particularly useful, but the greater the time lapse between each of the studies and the last Census (in the UK in 2011), the less current is this data source. The studies have also utilised, where available, commercial data and third-party data sources, often supplied by local authorities. Again, particularly in the case of the Welsh studies, local authorities have been exceptionally supportive in sharing data for inclusion within each study. The availability of quality data has tended to vary from place to place and has tended to be more extensive and readily available in Wales than in Australia and Vanuatu. As far as possible, data is drawn from the lowest available spatial scale so that a detailed profile of each community can be built. In the Welsh studies, for example, this has typically involved Lower Layer Super Output Area small area statistics. Where relevant data is not available at this spatial scale, the next largest scale where data is available has been used as a substitute.

Coproduction and 'Think Space'

Having undertaken the initial desk-based research components, the Deep Place case studies have then proceeded to more participative methods of enquiry. These have sought to go beyond the quantitative data to establish a complete understanding of the 'lived experience' of each community. The initial stage of this primary research has involved the mapping of agents and agencies. Typically, public sector organisations are engaged first, followed by third and private sector partners. Alongside this, and sometimes reflecting discussions with the social partners, the research team has sought to identify a representative sample of community activists and residents. The result of this mapping process is the construction of a panel of individuals, with representatives of agencies grouped into policy areas, as well as an overarching group of community representatives.

This panel of individuals is then invited to attend a series of 'Think Spaces', which essentially take the form of seminars of no more than 12 participants that are loosely structured around particular policy areas. Participants for each session are drawn from public-, private-, and third-sector organisations, sometimes including an academic associate with a particular policy expertise to help inform discussions, and, participating on an entirely equal basis, community representatives. Themes for each seminar, are informed to some extent by the findings of the socio-economic analysis that is conducted but typically include: economy; education and skills; housing; community safety; health and social services; environment; and overarching community sessions. If the recruitment process works well, the ensuing conversations are free-flowing and lead to informed views on actions necessary to help ensure the future sustainability of the community. Although the challenges faced by each community are not disregarded, the emphasis is placed on positive locally grounded and future-orientated opportunities informed by the Transition Thinking process of 'backcasting' we discussed above.

Members of the research team seek to facilitate, rather than steer each Think Space session. We have found that the choice of setting and even the layout of the room can have an important impact on the tone of the conversation, as can the number of participants, which is why numbers are limited to around 12. Meeting rooms at local authority offices, for example, are avoided in favour of more informal settings. In the past, sessions have been held at community centres, heritage attractions, rugby clubs, and indoor markets. We have found that such locations tend to help break down any barriers that exist between the community members and agency representatives that are participating. These sessions are supplemented with smaller, often one-to-one conversations with participants to explore points further. All discussions held throughout the research process, whether during the Think Spaces or as part of the supplementary process, are held under Chatham House Rules. This meant that even though these conversations have strongly influenced the writing of each report and the conclusions reached, contributions made during the research processes are never directly attributed to participants. This not only helps remove barriers to participation but also protects individuals and therefore assists free-flowing conversations.

Atmosphere Landscape Horizon and place check

The collation and presentation of data from the extensive research process initiated by a Deep Place study can present challenges of accessibility, particularly for lay audiences. Finding a coherent reporting framework is, therefore, important to help ensure the effective dissemination of findings and to assist in developing a local consensus on future actions. In some cases, a framework called Atmosphere, Landscape, and Horizon (ALH) has been utilised to help structure data reporting. Initially used to deliver a JRF review on the impact of devolution on place-based regeneration policy in the devolved regions of the UK (Adamson,

2010), the ALH framework was further developed by the Centre for Regeneration Excellence Wales (CREW).

The ALH framework identifies three related perspectives on the social experience of life in a community – atmosphere, landscape, and horizon – which were summarised by Adamson and Burgess (2012, p. 9) as follows:

> **'Atmosphere'** refers to the 'feel' of a place. Is it somewhere you would like to live? Is it welcoming or hostile? Does it feel safe or dangerous? Do residents have a sense of identity, and is that positive or negative? Is it a tight-knit community or one fractured by difference and inequality? Is it a community respected by others or stigmatised for crime, anti-social behaviour and substance misuse? All these issues are key parts of the regeneration process. When we regenerate a place, we make it somewhere that's good to live and where residents feel pride and a sense of belonging
>
> The **'Landscape'** of a place reflects its physical characteristics and the ways in which they influence the quality of life for those who live there. It is about the structure and design of the built environment and the public realm. What is the housing quality and its appearance? What is the housing density and distribution? Are there green spaces and gardens, and are they well kept and attractive? Are there play and sports spaces, and functional public spaces which contribute positively to the atmosphere of the community? Are there services required by residents for shopping, learning, health and well being, and exercise? Is the community well connected internally by good roads and paths, and externally by transport links and information technologies?
>
> The **'Horizon'** of a community describes the sense of social horizon experienced by residents. It is concerned with cultural and psychological horizons, and the ways in which residents orientate themselves towards the external world. Are they empowered to interact with the wider social and economic world, or is life restricted to the community by low educational attainment, worklessness and lack of confidence? Do residents travel outside for work, leisure and learning or are they trapped by low aspirations within a peer culture which is passive and lacks direction? Most fundamentally, is there social provision which builds bridges to the outside worlds of employment, education and healthy living?

These three perspectives combine internal community experiences with the external perceptions of a place that are held by wider society. In formulating responses to these perspectives, the framework engages researchers and community in an appraisal of physical and environmental factors, as well as social, psychological, and cultural experiences of life in a physical space. An associated ALH Toolkit (Centre for Regeneration Excellence Wales, 2012) identified a range of quantitative and qualitative indicators available in the UK, and similar indicators were identified for use in the Australian Deep Place study (as discussed in

Chapter 9). In the Vanuatu Deep Place study, however, comparable data sources were not available (also discussed in Chapter 9).

Gathering data for the ALH framework, and particularly the Landscape perspective, requires a physical review of the local environment, community facilities, and quality of the public space. A range of techniques have been developed for acquiring and recording this type of information, and the most successful of these facilitate participation by local communities and help capture their lived experiences[1]. Each of these methods has the capacity to inform the physical appraisal of place and to provide qualitative data for the ALH approach. The Deep Place studies adopting the ALH framework have tended to adopt the Place Check technique, largely because of its accessibility, ease of delivery, and limited financial costs. Whichever technique is used, however, the principles of democracy, accessibility, and participation are fundamental. Ultimately, the technique should empower the community to envisage positive change and co-produce the solutions and actions that emerge in response to the findings of the Deep Place exercise.

Horizon scanning

The UK Government Office for Science's Futures Toolkit (2017) sought to provide guidance on how policymakers could embed long-term thinking in the making process. The toolkit identified horizon scanning as an important technique for helping policymakers look ahead, consider the future rather than the present, and identify issues that will be important that are likely to be different from those of the present. It identified three horizons: short term, medium term, and long term. Identifying and mapping the drivers of change (the key trends shaping the long-term development of a policy area) are at the core of this kind of futures work. Various horizon scanning methods have been utilised. The Food and Agriculture Organization of the United Nations (FAO, 2013), for example, identified several different techniques, including best-worst scanning for prioritising trends or developments; delta scanning for capturing identified trends and developments from other horizon scanning processes; expert consultations for tapping specialist knowledge; and manual scanning to identify signals of change to track trends and drivers.

Ultimately, as the UK Government Office for Science highlighted, horizon scanning exists in an area of policymaking that lies outside of a firm evidence base and tends to rely to a greater extent on the instincts of the team involved. Horizon scanning is, therefore, able to explore innovative and unexpected issues, as well as persistent problems and previous trends. The Deep Place case studies have utilised this horizon scanning technique to focus on places, both with a view to considering the overall impact medium- and long-term drivers for change on the settlement, but also in considering how changes in specific policy areas might affect the communities. To some extent, this is reflected in the overall approach taken by the research team in undertaking the research. The purpose of each study has been to consider how particular communities can become more sustainable over the medium term to long term but is also embodied in the particular research techniques utilised during each case study. The structure and format of the Think Spaces, for example, are very much envisaged as horizon scanning methods and embody Transition Thinking backcasting techniques. The literature reviews undertaken, and the illustrative case studies included in each report are designed to identify social and economic innovation.

Action points and Coalition for Change

Each study has presented a range of action points that are collectively aimed at moving that community towards a more economically, socially, culturally, and environmentally sustainable model over the medium to long-term. Many of these have been relevant to particular communities, whilst others are more broadly applicable, albeit with a high degree of locally specific application. Action points are a direct result of the evidence presented during each case study. Evidence is drawn from each of the different methods that are deployed, with a strong emphasis on discussions during each of the Think Spaces. Illustrative case studies are used in each Deep Place report to provide examples of similar changes to those advocated in each of the action points that are being undertaken elsewhere, thus avoiding speculative recommendations that are not based on working examples. Ultimately, of course, there is some degree of interpretation, and an inevitable degree of bias, on the part of the research team. This bias is inevitably based on our shared perspectives as presented in Part One. Each of the action points is, however, evidence both from the primary research undertaken as part of each study and from the literature reviews.

Whereas each of the action points has identified a particular lead agency that would be responsible to take forward the change advocated, following the first Tredegar case study it became apparent that there needed to be a forum to take broader ownership of the research programme as a whole, following the formal departure of the research team. Initially, during the dissemination process following the publication of the Tredegar study, it was hoped that such a structure would emerge organically from those who had actively participated during the primary research phase. This proved not to be the case, however, partly because there were certain changes in key personnel at various agencies in the months that followed. In subsequent studies, a new methodological development was advocated. Consequently, every study since Tredegar has promoted the creation of a 'Coalition for Change' to take forward, although not be limited to, the overall implementation of the research report and chart progress. The Coalition for Change structure is designed to ensure lasting leadership and provide a forum for partnership focused on the future sustainability of each community. Conceived on the same basis as the coproduction principles of the research programme itself, the Coalition for Change forum is designed to be built from, though not limited too, the agents and agencies that were engaged during the Think Spaces. The Coalition for Change must continue to adapt, evolve, and be open to further and greater involvement.

Summary

This chapter has outlined the methodological influences and specific research methods utilised during the various Deep Place studies that have been completed, and which are summarised in Chapters 8 and 9 that follow. This second part of this book has been designed to offer a more practical presentation of how we have sought to progress our wider theoretical discussion in earlier chapters. Whilst the two chapters that follow provide a summary of the Deep Place case studies we have undertaken in various communities in Wales, Australia, and Vanuatu, this chapter has outlined our specific methodological influences in doing so and summarised the methods we have used in undertaking these case studies. These methods have not been rigidly adhered to in each study, although the general approach has remained consistent, as they have been adapted to the specific and unique context of each study. A strength of the approach is its ability to recognise local conditions and reflect them in the mix of methods deployed, and to reflect our knowledge of what works in undertaking this kind of research as it has evolved.

No research is conducted without some degree of methodological influence, and in this chapter, we have sought to codify the main influences that have impacted our approach. Thinking around social exclusion, transition theory, Total Place, the Foundational Economy and anchor institutions, and more general sustainable place-making, have been most influential. The references provided in this chapter, as indeed elsewhere in this book, are designed to provide a wider reading list that the reader may wish to consult as a next step. The research methods outlined in this chapter, whilst not designed to be a strict instruction manual that *must* be followed, do offer options for those wishing to undertake the kind of 'Deep Place' planning we have espoused in this book. What is important, however, is that the inclusive, co-productive, and whole-place principles underpinning our research are embedded in any sustainable place-planning agenda.

Our research is experimental and constantly evolving, with different aspects being amended, added, or omitted from place to place. As shown in the case studies that follow, every place is different and therefore the research process needs to flexible and adaptive. As we progress further into Deep Place case studies in the future, our own approach is likely to evolve further. If we were to honestly appraise our research, we would admit to varying degrees of success. In some cases, many of the action points have been or are actively being progressed. In other cases, rather fewer have been taken forward. In no case have all our action points been enacted, but we never expected them to be. What we have sought to do is challenge local and regional thinking about what a sustainable future looks like for individual communities, and how to move towards such a future. Nor were timescales necessarily designed to be immediate, the Deep Place studies have specifically been written as multi-generational in their scope. We have sought in the Deep Place studies summarised in the following chapters to suggest to communities and local agencies an alternative path, and the decision to take that path is an important first step.

Note

1 See, for example, Planning for Real (https://www.planningforreal.org.uk); Community Star[TM] (https://www.outcomesstar.org.uk); PhotoVoice (https://www.photovoice.org); and, Place Check (https://placecheck.info/en/).

References

Adamson D (2010) *The Impact of Devolution: Area-Based Regeneration Policies in the UK.* York: Joseph Rowntree Foundation.

Adamson D and Burgess K (2012) *Atmosphere, Landscape and Horizon: A Regeneration Assessment Toolkit.* Merthyr Tydfil: CREW.

Augar P (2019) *Independent Panel Report to the Review of Post-18 Education and Funding.* London: MH Government.

Bentham J, Bowman A, de la Cuesta M, Engelen E, Ertürk I, Folkman P, Froud J, Johal S, Law J, Leaver A, Moran M and Williams K (2013) *Manifesto for the Foundational Economy.* Manchester: Centre for research on Socio-Cultural Change, University of Manchester.

Birley A (2017) *6 Steps to Build Community Wealth: Using What We Already Have to Generate Local Economic Growth Co-operatively.* London: Co-operative Party.

Blumenfeld H (1955) 'The economic base of the Metropolis: Critical remarks on the "basic – nonbasic" concept', *Journal of the American Institute of Planners*, 21(4): 114–132.

Bramley G, Hirsch D, Littlewood M and Watkins D (2016) *Counting the Cost of UK Poverty.* York: JRF.

Castells M (1996) *The Rise of the Network Society: The Information Age: Economy, Society and Culture, Vol 1.* Cambridge, MA: Blackwell.

Centre for Regeneration Excellence Wales (CREW) (2012) *Atmosphere, Landscape, Horizon Toolkit Questionnaire.* Merthyr Tydfil: CREW. Accessed 18 November 2021: http://www.regenwales.org/resource_17_The-Atmosphere--Landscape-and-Horizon-Regeneration-Impact-Toolkit.

De Boeck S, Bassens D and Ryckewaert M (2019) 'Making space for a more foundational economy: The case of the construction sector in Brussels', *Geoforum*, 105: 67–77.

FAO (2013) *Horizon Scanning and Foresight: An Overview of Approaches and Possible Applications in Food Safety.* Rome: Food and Agriculture Organization of the United Nations.

Francis-Devine B (2021) *Poverty in the UK: Statistics.* London: House of Commons Library.

Froud J, Johal S, Moran M, Salento A and Williams K (2018) *Foundational Economy.* Manchester: Manchester University Press.

Grint K (2009) *Total Place: Interim Research Report: Purpose, Power, Knowledge, Time and Space.* Warwick: University of Warwick Business School.

Grint K and Holt C (2011) 'Leading questions: If "Total Place", "Big Society" and local leadership are the answers: what's the question?', *Leadership*, 7(1): 80–98.

Hansen T (2021) 'The foundational economy and regional development', *Regional Studies*. https://doi.org/10.1080/00343404.2021.1939860

Harkavy I and Zuckerman H (1999) *Eds and Meds: Cities' Hidden Assets.* Washington, DC: The Brookings Institution.

Heinberg R (2008) 'Foreword', in: R. Hopkins (ed.) *The Transition Handbook: From Oil Dependency to Local Resilience.* Totnes: Green Books.

Hodges R and Dubb S (2012) *Road Half Travelled: University Engagement at a Crossroads.* East Lansing: Michigan State University Press.

Jackson B (2019) 'The power of place in public leadership research and development', *International Journal of Public Leadership*, 15(4): 209–223.

Just Transition Commission (2020) *Interim Report*. Edinburgh: Scottish Government.

JRF (2013) *Monitoring Poverty and Social Exclusion in Wales: 2013*. York: JRF and New Policy Institute.

JRF (2020) *UK Poverty 2019/20: The Leading Independent Report*. York: Joseph Rowntree Foundation.

Lang M (2020) *Can You Get There from Here? Post-16 Education, Social Progression and Socio-Economic Resilience*. Tongwynlais: ColegauCymru CollegesWales.

Leadership Centre for Local Government (2010) *Total Place: A Practitioners' Guide to Doing Things Differently*. London: Leadership Centre for Local Government.

Milbourne P (2011) 'Poverty in rural Wales: material hardship and social inclusion', pp. 254–270 in: P Milbourne (ed.) *Rural Wales in the Twenty-First Century: Society Economy and Environment*. Cardiff: University of Wales Press.

North P and Longhurst N (2013) 'Grassroots localisation? The scalar potential and limits of the transition approach to climate change and resource constraint', *Urban Studies*, 50(7): 1423–1438.

Sachs J, Kroll C, Lafortune G, Fuller G and Woelm F (2021) *Sustainable Development Report 2021: Includes the SDG Index and Dashboards. The Decade of Action for the Sustainable Development Goals*. Cambridge: Bertelsmann Stiftung, Sustainable Development Solutions Network and Cambridge University Press.

Taylor H and Luter G (2013) *Anchor Institutions: An Interpretive Review Essay*. New York: Anchor Institutions Task Force.

UK Government Office for Science (2017) *Futures Toolkit*. London: Government Office for Science.

United Nations (2016) *Leaving No One Behind: The Imperative of Inclusive Development*. New York: United Nations.

United Nations (2017) *New Urban Agenda*. A/RES/71/256. New York: United Nations.

United Nations (2020) *The Sustainable Development Goals Report 2020*. New York: United Nations Publications.

United Nations General Council (2015) 'Transforming Our World: The 2030 Agenda for Sustainable Development', *A/res70/1. Resolution Adopted by the General Assembly*. Accessed 18 November 2021: https://www.un.org/ga/search/view_doc.asp?symbol=A/RES/70/1&Lang=E

8

CASE STUDIES

UK

Introduction

In the UK, four Deep Place studies have thus far been completed: Tredegar (Adamson and Lang, 2014), Pontypool (Lang, 2016), Lansbury Park (Adamson and Lang, 2017), and Llandovery (Lang, 2019). Although based on the authors' previous experience and expertise, as the first study, Tredegar was to some extent experimental. This led to some interesting methodological developments that have influenced each of the subsequent studies. We hoped that a further study, of the Cardiff suburb Trowbridge, would be completed in 2021, but due to the COVID-19 pandemic and the restrictions this placed on undertaking research of this kind, only the initial desk-based socio-economic analysis has been completed. We hope that this study will be undertaken at a later date. Further studies have also been undertaken in Australia and Vanuatu, and these are discussed in Chapter 9.

Located at the top of the South Wales Valleys, Tredegar is in Blaenau Gwent, which is one of the UK's most deprived counties. The town owes its modern existence to the rapid rise of industrialisation during the 19th century, which, because of employment opportunities created firstly by iron production and then coal mining, attracted migrant labour largely from elsewhere in the UK. Between 1801 and 1881, Tredegar's population rose from 1,132 to 34,685 (Powell, 1884). The comparative isolation of the settlement and the hardships experienced by working families led the community to forge strong political and social values of mutual aid and socialism. Tredegar's Medical Aid Society, funded by weekly small contributions from the workers' wages, influenced Tredegar-born Aneurin Bevan who later used it as the model for the UK's National Health Service (Davies, 2007). During the post-war period, however, the town began a long

DOI: 10.4324/9781003221555-10

process of post-industrial decline, and its sense of isolation was amplified as the levels of deprivation grew.

Pontypool is located in the middle of the Afon Lwyd Valley, the most easterly of the South Wales Valleys. The town has a persuasive claim to have been the first industrial town in Wales and it became an important location for the production of so-called 'japan-ware' (tin-coated iron sheets). The first forges and furnaces were built in nearby Pontymoile in the 15th century, with further significant development following in the late 17th century, whilst heavier industries came to dominate the local economy during the 19th century. The town grew steadily throughout the 19th century, and a town hall was built in 1856, while an Urban District Council was established in 1895. Like Tredegar, however, post-war industrial decline has been a significant factor in the town's more recent experiences. Nevertheless, during the 20th century, the town continued to grow and there were substantial public-housing schemes in the suburbs of Pontypool. The town now has a population of around 23,400 people with extremely mixed socio-economic patterns.

Lansbury Park is a large post-war local authority housing estate on the outskirts of the Caerphilly town centre. The community has experienced significant levels of poverty and social exclusion over a prolonged period. The estate had appeared in Adamson's (1996) study of poverty in Wales, which found around 50 percent of working-age males of Lansbury Park at that time were not in employment, and that there were very high levels of lone parenthood. These characteristics, Adamson argued, caused complex patterns of social exclusion that were triggered by poverty and worklessness. Today, over 25 years later, Lansbury Park continues to experience similar patterns of social exclusion that have now become embedded at a socio-cultural level. Despite numerous interventions, the level of poverty in Lansbury Park remains among the highest in the UK and social exclusion continues to be the dominant experience for residents. At the time that study was undertaken, according to the Welsh Index of Multiple Deprivation, Lansbury Park was the most deprived community in Wales.

Llandovery is located in rural Carmarthenshire in West Wales. It lies around 30 miles north-east of Carmarthen on the edge of the Brecon Beacons National Park and is one of the county's market towns. The town traces its earliest history as a settlement to a Roman fort built near the town, but the modern settlement was established after the construction of the Norman Llandovery Castle. Llandovery quickly developed the role of a drovers' town as it is located at the intersection of three historically significant droving routes. This cemented its strong links to the agricultural community and its provision of facilities as a rural market town. Llandovery, for example, saw the establishment of the Black Ox Bank in 1799 which was initially developed to provide banking facilities for drovers and the agricultural community. Today, Llandovery has a population of around 2,600 people.

Tredegar was chosen as the location of the first study following desk research that had initially identified six potential locations. Of these, Tredegar was found to be the most appropriate as it was compact enough to have a meaningful depth of area-based analysis but also appeared to experience a broad range of challenges and offer multiple opportunities on which a successful sustainable place-making transition could be initiated. Initially, the study area consisted of just the core part of the town. In discussions with the community, however, it quickly became apparent that two neighbouring areas (Sirhowy and Georgetown) should be added to the study, as they were generally considered by residents to be connected to the town. As discussed in the previous chapter, this co-productive boundary-setting approach has proved to be a typical experience in each of the studies. Llandovery was originally envisaged to be a parallel study to be undertaken alongside

FIGURE 8.1 Map showing the location of the UK case studies.

Tredegar to test the methodology within a rural context. Due to administrative complications, an opportunity to undertake that study arose only later.

Pontypool was selected as an appropriate location to further test and refine the methodology established in Tredegar, as although it shared share some of Tredegar's local challenges and opportunities, it was sufficiently different to offer some additional understanding of sustainable place-making. The most striking difference between the two, which became clear during the initial desk-based selection process, was the variation of relative deprivation that existed in the town. Tredegar's experience had been one of almost universal disadvantage in comparison to a more differentiated social structure in Pontypool. Lansbury Park differed from the other studies as it was formally requested by its local authority, Caerphilly Council. The council wanted an innovative study into the causes of the entrenched poverty experienced on the estate with recommendations on how to overcome them. Lansbury Park also differed from the other studies as it incorporated an additional methodology, the Atmosphere, Landscape, Horizon framework, which was discussed in the previous chapter.

National and regional context

Wales has experienced a prolonged period of economic decline that, despite over 20 years of devolved governance and considerable European Regional Development Fund investment, continues unabated. The Welsh economy has had the lowest productivity rate of the four UK nations or any region of England, and Gross Value Added (GVA) per head in Wales has varied between 70.5 percent and 74.4 percent of the UK average since devolution (Munday et al., 2018). In those areas where Deep Place studies have been undertaken, West Wales and the South Wales Valleys, GVA per head is significantly below the Welsh average. The Gross Disposable Household Income per head in Wales in 2018 was £17,100, which was lower than each of the other UK nations and English regions (ONS, 2020). Gross Average Full-Time Weekly Earnings is the standard measure to compare earnings from work, and, in April 2017, Wales had the lowest average weekly earnings amongst the four UK countries and English regions at £498 (IPPR North, 2018).

The number of jobs in Wales increased by 26.7 percent between March 1999 and March 2018, whereas over the same period the number of jobs across the UK increased by 20.3 percent (Welsh Government, August 2018). Public sector employment as a percentage of total employment in Wales in 2016 was 22.4 percent (IPPR North, 2018). In 2017 there were 250,100 active businesses in Wales, with a combined annual turnover of £117bn. 95 percent of business in Wales are micro-businesses employing nine people or fewer, whereas large businesses account for 0.7 percent of the total, but employ 38 percent of the workforce. Medium-sized enterprises, those employing between 51 and 249, represent less than 1 percent of SMEs in Wales. (Munday et al., 2018) Despite its large micro-business base, and its so-called 'missing middle' of locally grounded medium-sized firms, much of the focus of Welsh economic policy throughout the period of devolution has been on foreign direct investment (FDI), seeking to attract large externally owned firms into Wales. This strategy predates devolution and was pursued actively by the Welsh Development Agency until its absorption into Welsh Government in 2006 where FDI has remained a core strategy.

Although there have been more recent experiments with the so-called 'Foundational Economy', overall Welsh economic policy has given priority to sectors with the greatest perceived growth potential. Enterprise zones have been established and City Deals (an economic programme of the UK Government designed to support regional growth) have been signed and supported by the Welsh Government and Welsh local authorities (Lang and Marsden, Forthcoming). Significantly over half of the financial investment envisaged as part of the Cardiff Capital Region City Deal has been earmarked for improving sub-regional physical connectivity via a so-called Metro which, it is envisaged, will be an 'engine of economic growth'. The Metro has been founded on a conviction that public investment should be based on opportunity rather than need as, it says, 'a philosophy which has concentrated investment on need rather than opportunity, has resulted in an economy plagued by homogeneous mediocrity', and it states that directing resources at the South Wales Valleys has not worked (Barry, 2011, pp. 4–5). This appears to be consistent with the agglomeration logic discussed in Chapter 2.

The UK Government accepted the Committee on Climate Change's (2019) advice to set a net-zero target for GHG emissions by 2050, and both the Welsh and Scottish governments have set similar targets. It should be noted, however, that the 2050 target only relates to activities taking place within the UK's borders (territorial emissions). This is important to note, as during 2016 the UK's consumption emissions were around 56 percent higher than territorial emissions. The UK's devolved administrations have important roles to play in tackling climate change and, on a percentage basis, in 2017 territorial emissions in Wales fell by 13 percent, whilst Scotland and Northern Ireland's territorial emissions both fell by 3 percent. Territorial emissions in England fell by 1 percent. It should be noted, however, that the reductions in territorial emissions across the devolved administrations in 2017 were almost entirely a result of changes in the power sector, which Scotland and Wales have little control over. As some important policy areas have not been devolved, the ability of devolved administrations to pursue their own totally consistent approaches is to some extent limited. Nevertheless, Wales has agreed to potentially groundbreaking legislation to safeguard future well-being.

The Well-being of Future Generations (Wales) Act (2015) has placed a legal duty on all devolved public bodies in Wales to protect the economic, social, environmental, and cultural well-being of Wales. Based on a Sustainable Development principle, the Act contains seven Well-being Goals:

- Prosperous Wales,
- Resilient Wales,
- Healthier Wales,

- More Equal Wales,
- Wales of Cohesive Communities,
- Wales of Vibrant Culture and Thriving Welsh Language, and
- A Globally Responsible Wales.

The Act also requires what it calls 'Five Ways of Working', which devolved public bodies must focus on long term, integration, involvement, collaboration, and prevention. Important though the legislation is, however, its impact has not yet been fully felt. The Welsh Future Generation Commissioner's Office has, for example, asked: 'can the whole infrastructure of [Welsh] Government answer the question of whether its spend increases or decreases carbon emissions in Wales? From what I have seen, the answer to that question is no' (IWA, 2021).

The national and regional context within which the Deep Place studies have been undertaken, therefore, is one where the Welsh economy continues to experience structural difficulties. Despite some recent marginal experiments with the 'Foundational Economy', the emphasis in Welsh economic policy remains one overwhelmingly focused on an elusive pursuit of growth, through an emphasis on agglomeration and FDI. Moreover, although there have been improvements in emissions, these have been largely dependent on changes in the power sector. Wales has passed groundbreaking Future Generations legislation, but as the Commissioner's Office highlights, there are particularly strong tensions with existing Welsh economic policy, and the success of programmes such as the non-devolved City Deals, for example, are measured on their contribution to growth rather than well-being. As we discussed in Chapter 2, well-being and growth are not necessarily compatible. Each of the studies has sought to understand the relevance of this broader context for each of the communities studied.

Challenges explored and windows of opportunity

Each of the UK Deep Place studies has identified a range of challenges that currently impact each community. The studies have sought to understand these challenges as part of an overall approach to identify the actions necessary to secure future economic, social, environmental, and cultural sustainability. The major challenges consistently evident in each community have included: experiences of deprivation, health inequalities, education and skills attainment gaps, poor quality housing, and limited public transport provision. The impact of each of these varies between and within each of the communities.

Health

Health inequalities have been a significant factor in the Deep Place studies, and the effects on the communities studied have been clearly detectable. The Tredegar study identified a 20-year healthy life expectancy gap between those living in the most and least disadvantaged communities the Aneurin Bevan Local Health

Board (LHB), which besides Tredegar, also contains Pontypool and Lansbury Park (Llandovery is located in the Hywel Dda LHB in West Wales). Across Wales as a whole, the healthy life expectancy gap is 19 years for men, and 18 years for women. Public Health Wales and the Welsh Government (2016) have recognised that the extent of this inequality is repeated in many areas of health. They have found that the incidence of cancer is 23 percent higher, and cancer mortality is 48 percent higher, in the most deprived areas when compared to the least deprived areas. Furthermore, 14 percent of all reception year children living in the most deprived areas of Wales are obese, compared to 9 percent in the least deprived areas (p. 2). When published, the Welsh Government's (2011) health inequalities action plan demonstrated that the inequality gaps in health and well-being had been increasing over the previous 20 years.

Health professionals in Torfaen identified that the key issues affecting the health of the residents of Pontypool at the time that study was undertaken included: low-level mental health needs; the effects of poor housing conditions on health; the effects of financial stress; and social isolation and loneliness (typically affecting the over 60s, but in poorer areas affecting the over 30s). These issues present as both mental and physical ill-health. The Fabian Society (2009) suggested that people living on estates such as Lansbury Park were twice as likely to suffer from mental health problems as the general population. It should be noted that in Llandovery, however, public health indicators appeared to demonstrate that the general population at the time of the study tended to compare well to wider Welsh averages. Discussions held during that study indicated that this may be explained, in part, by particularly active local lifestyles, as well as partly by the comparatively much cleaner environment in the rural area.

The Marmot Review (2010) identified that health inequalities do not arise by chance and are the result of social and economic circumstances. We would suggest environmental circumstances are another major factor. Marmot suggested that traditional public health campaigns had little effect on changing behaviours and that underlying patterns of consumption are primarily caused by societal inequalities. Moreover, Julian Tudor Hart's (1971) Inverse Care Law argued that the availability of good medical care tended to vary inversely with the needs of the population served. The Tredegar, Pontypool, and Lansbury Park studies argued that public health policy needed to be underpinned by comprehensive anti-poverty strategies, and by the need to ensure the effects of disadvantage on access to services and outcomes were minimised.

Although accessing local health services is certainly an important consideration in tackling health inequalities, each of the Deep Place studies has argued that if the underlying causes of these inequalities are to be addressed, there must also be a strong focus on local economies. Harnessing the potential local economic benefit of health service expenditure has also been identified in each of the studies. The Tredegar study, for example, highlighted the work of the Evergreen Cooperative Initiative in Cleveland, Ohio, which was launched in 2008 by a group of 'anchor institutions' to create living-wage jobs in six low-income neighbourhoods. One successful initiative was a company that grows lettuce and herbs for commercial sale, including to the anchor institutions that helped set up the initiative. The benefits of this kind of local procurement have been emphasised in each of the studies.

Education and skills

The Deep Place studies have identified significant attainment gaps in the respective communities. These attainment gaps between children from deprived backgrounds and those from more affluent families grow significantly between ages 7 and 15. This is symptomatic of a wider trend across the UK, where children from the lowest income families are half as likely to get five good GCSEs (General Certificate of Secondary Education) and study subsequently at university (Sharples et al., 2010). David Egan (2013a, 2013b) has suggested that part of the reasons for these inequalities arises from the fact that pupils from poor backgrounds do not have sufficient parental support for their education from the age of seven onwards. This is not, he says, because these parents do not want their children to do well, but because they do not have the ability to support them or to assist with homework. He also suggested that as children get older, the pressure of their peer groups in poor communities also becomes a limiting factor in their educational achievement. There are, of course, a wide range of other factors that impact attainment gaps, such as family financial inability to continue to support children beyond compulsory education, domestic violence, and family physical or mental health issues that turn children into young carers.

The studies have found that free school meal entitlement is a powerful indicator of the local barriers to educational attainment. The Tredegar study identified that just 12 percent of pupils qualifying for free school meals leave education with five good GCSEs that include Mathematics and English, and only 16 percent of 18- to 19-year-olds enter higher education, compared with 57 percent in the least deprived areas of Wales. The Lansbury Park study found a clear and long-term profile of free school meal entitlement in that community, which had over twice the national average entitlement with some 56 percent of pupils receiving free school meals. Consequently, that community has also been characterised by lower attainment rates. The studies have also found that pupil absence is another powerful indicator of likely attainment rates, and in Pontypool, this was an acutely notable challenge.

Schools in rural areas face particularly severe challenges resulting from the rationalisation of provision. The Llandovery study was undertaken shortly after the closure of the town's secondary school, and although the community appeared realistic about the unlikely reopening of the school, it was still bitterly regretted. The decision to close the secondary school was taken because of declining pupil numbers, which is symptomatic of longer-term trends of rural depopulation, particularly of younger people. Although the study found a strong degree of local confidence in the future of Llandovery's primary school, which draws pupils from a wider geographical area, in the long term its continued provision will depend on maintaining thriving communities in and around Llandovery.

Despite the closure of the secondary school, there is a real commitment to make meaningful use of the former school's premises, and Carmarthenshire Council announced plans to move Llandovery's primary school onto the site following a refurbishment programme. The new school premises was also likely to have surplus space that would be available for wider community use, and this would appear to offer an ideal facility for wider community education. This would also accord with a model of community-focused education, encompassing: highly motivated and professionally run schools that refuse to accept that poverty should lead to low achievement; the close involvement of parents and families in the education of their children; and the local community engaged in a meaningful relationship with schools (Egan, 2013a). This model of community-focused schooling has been supported in each of the Deep Place studies.

Each of the studies has also been concerned with the nature of skills taught and the methods by which they are advanced. The studies have found that the true potential of vocational education has not been fully realised and there tends to remain a pervasive emphasis in Wales on GCSEs and A Levels as the gold standard. The studies have therefore suggested that a greater emphasis on further and vocational education was needed, as this would help address skills shortages and overcome the attainment gap. In this respect, when the Tredegar study was published a major opportunity had arisen with the opening of the nearby Coleg Gwent Learning Zone campus in Ebbw Vale, to provide all post-16 educational provision in the area. Similar plans were being developed in Torfaen when the Pontypool study was published, and a new facility in nearby Cwmbran has recently opened. As the Llandovery study was being undertaken, a Rural Skills and Innovation Hub was being developed in Carmarthenshire. These positive developments in post-16 education are significant for the improvement of future educational attainment.

Housing

The availability of good quality and affordable housing is one of the most basic human needs, and it has therefore formed an important consideration of each of the Deep Place studies. Moreover, the provision of housing services such as construction, repair, renovation, and low carbon retrofitting also provides considerable opportunities to promote local economic development. In Wales, the advent of devolution led the way to significant changes in the structure of social housing. At the time, local authorities in Wales were struggling to maintain their housing stock because of 20–30 years of underinvestment from the central government. Improving social housing became a major priority for the new devolved government, which introduced a Welsh Housing Quality Standard (WHQS).

Local authorities began to focus on options to meet the standard and the idea of transferring council housing into housing associations, which would be able to borrow to fund the necessary investment, emerged. The community mutual model, which ensured tenant participation, became the preferred option in many cases. Consequently, the majority of councils held tenant ballots on the question of transfer, and 11 new housing associations were created across Wales as a result. Meeting WHQS was a key part of the offer made to tenants in the ballots that were held, and most of the new associations had met the terms of their offer within about five years. Caerphilly was one of the local authorities that did not transfer, and the local authority continues to manage and maintain its own housing stock. A major investment by the local authority in housing quality was already well underway at the time the Lansbury Park study was undertaken. Whether through new housing associations or local authorities, therefore, investment because of WHQS is helping to ensure that social housing in Wales is no longer the last choice in the housing market.

Some parts of Tredegar, Pontypool, and Llandovery had high levels of owner occupation, but these studies found that this was not a guarantee of good quality housing. The traditional housing stock of the South Wales Valleys and large parts of rural Wales are ageing and, in many instances, in a poor state of repair. An ageing population coupled with low incomes also suggested some critical barriers to future improvement. Meanwhile, private-sector rented accommodation has historically produced some of the poorest quality accommodation in the housing market, and heavy concentrations of private rental properties were identified in parts of Tredegar and Pontypool. These studies suggested that engaging with landlords in a housing renewal approach or in retrofitting low carbon measures can be difficult, and they consequently argued for a robust approach to ensure that the private-sector rented stock does not become a significant problem in achieving these wider aims.

As well as the quality of existing housing, the studies also addressed the critical issue of the supply of affordable homes. Lack of access to affordable housing in rural areas, as discussed in the Llandovery study, is widely regarded as a significant symptom of rural poverty. In areas with a strong housing market, affordable housing is usually achieved on the back of private-sector building programmes. The difficulties in Tredegar and other communities in the South Wales Valleys are exacerbated by the fact that volume house builders do not regard them as 'aspirational' localities and do not anticipate sufficient market interest to merit development. Additionally, local market prices tend to prevent conventional profit margins from being achieved and the level of risk is seen as too high. As a result, there is effectively a 'snowline' above which these house builders are not prepared to venture into the South Wales Valleys. Whereas Tredegar is well above this snowline, Pontypool sits on it and consequently, there has been some degree of recent private-sector development near the town.

Nonetheless, there has been an increasing interest in Wales in developing cooperative housing. The Calon Cymru Network, for example, seeks to develop sustainable rural communities across the Heart of Wales Railway and is aiming to increase the supply of affordable homes to be developed by a new Community Land Trust (CLT). The Llandovery study explored the use of Community Property Trusts (CPT), which are not-for-profit organisations that can be established by local people, to develop additional affordable housing supply in the town.

Each of these models has the capacity to provide long-term affordable housing, but, as the Llandovery study found, although community trusts often only provide small numbers of units, the process of establishing a CLT or CPT and obtaining planning permission is often slow, especially if the proposed development is regarded in some way as 'innovative' or outside the current local development plan (Powell et al., 2018).

Transport

Whereas more localised employment and entrepreneurial patterns have been consistently espoused by each of the Deep Place studies, they have also recognised that a certain degree of transport is clearly likely to remain a requirement. As the Llandovery study found, transport is a particularly significant issue for rural communities that are often most poorly served by public transport. Although Llandovery has a train station, which many rural communities have lost, train and bus services are infrequent and tend to not provide viable commuting options. The problems associated with infrequent and limited services are not, of course, limited to rural communities.

Tredegar has no train station and even though Pontypool and Lansbury Park do have local train services (in varying degrees of distance), these can be prohibitively expensive to use for many families. Each of the studies has also found that bus services were infrequent, time consuming, and, in some cases, expensive for those on lower incomes. Unsurprisingly, therefore, each of the studies found that private car transport was the most common form of transport relied on for commuting. Nearly 55 percent of Lansbury Park residents, however, had no access to a van or car, and this represented a major barrier to personal mobility and accessing activities that encouraged people from that community to experience other local cultures, as well as accessing employment and public services.

It should be noted that, in general, the average UK journey time to work has been steadily increasing. The National Travel Survey shows the average commute takes 28 minutes (DfT, 2011), but there is a close correlation between the commuting distance travelled per person per year and levels of household income. Individuals in the highest income quintile travel almost eight times further (2,529 miles) for commuting per year than those in the lowest quintile (DfT, 2011). This would appear to be broadly in accord with the patterns identified in the Llandovery study. The long-distance nature of rural commuting also makes other forms of transport, such as walking and cycling, an unlikely option for most who do not work locally. When accessing services or employment, people are prepared to cycle within a radius of about five kilometres, and the distance decreases even more when it comes to walking. Transport, and particularly public transport, is clearly also an important consideration when planning the provision of local services, but one which is often not considered during planning stages. Community transport often fills accessibility gaps in public transport, but it is not an inexpensive transport option and unit costs can often be higher.

The challenges identified in each of the communities studied have coalesced into similar themes, but their impact has varied significantly. Whereas Tredegar, Pontypool, and Lansbury Park experience significant public health weaknesses, Llandovery is comparatively far healthier overall. Educational attainment gaps are symptomatic of poverty wherever it is found, but some contributory factors such as attendance rates were more noticeable in some communities than others. Rationalisation of provision at post-16 offers opportunities to improve attainment, as in Tredegar, but presents wider community challenges in other areas such as Llandovery. Each of the studies found that WHQS was having an impact on improving social housing, but the private rental sector, particularly in Tredegar and Pontypool, offered more of a challenge. Affordable housing provision was a major issue in Llandovery, which, along with Tredegar, also experienced particularly notable difficulties in accessing affordable and reliable public transport.

Sustainable local economic activity

An emphasis on sustainable and local economic activity has been a central consideration for each of the Deep Place studies. These activities, we suggest, arise from the identification of goods and services that are required wherever communities are located. Food production and supply, energy conservation and generation, and health and social care have been consistent areas of economic activity considered in each study. As the studies have demonstrated, these goods and services offer significant potential to support a successful transition to sustainable and locally resilient communities, particularly as many of the critical assets necessary to support them are already found, to varying degrees, within or around these communities.

Food production and supply

Although the purchase of food represents a significant element of household expenditure, much of this spending is on supermarket chains. The Llandovery study, with evidence drawn from a recent local retail study (Nathaniel Litchfield and Partners, 2015), identified that the town had lost its share of the local food retailing market since 2008 (down from 50 percent to 44 percent), and, consequently, that it was now less self-sufficient in food retailing. This study found that none of the principal stores where household main grocery shopping was undertaken were independent, and only one of the principal stores was in Llandovery itself (albeit on the outskirts of the town). Furthermore, just 6.2 percent of Llandovery's households undertook 'top-up basket shopping' for groceries in Llandovery's independent shops. Although these issues were particularly notable in Llandovery, other communities face similar trends, as was notable in Tredegar.

As we discussed in Chapter 4, this pattern of food distribution and associated long supermarket supply chains has led to an increasing concern about food security and food quality. A more localised food economy, however, will require a major shift in consumption patterns, alongside radically altered supermarket supply mechanisms. The Tredegar study conceded that it is difficult to see either of these happening at scale without government intervention and wider societal recognition of the need for change. Beyond these broader systemic difficulties, there remain some significant barriers to local food production, particularly, as identified in the Tredegar and Pontypool studies, in the South Wales Valleys.

High rainfall and low hours of sunshine suggest that horticultural production in the Welsh Valleys would need to be undertaken under glass, but this would imply significantly increased costs of heating such facilities. Highlighting unpublished research undertaken by the State University of New York in partnership with the Centre for Regeneration Excellence Wales, the Tredegar and Pontypool studies urged careful consideration of the establishment of a product cluster based on local willow cropping, which amongst other uses could provide a virtually free supply of bio-mass heating. A further issue relates to the local food market. The communities of the South Wales Valleys are amongst the poorest in the UK and generally have lower disposable household incomes. Where there have been successful local food projects in the UK, they have tended to be close to large affluent markets. In these areas, middle-class consumer interests have permitted the development of local organic and other niche products, as a higher general disposable income is able to absorb additional costs. As indicated above, however, a close association with anchor institution procurement may help overcome some of these difficulties. Therefore, the growing of crops such as salad under glass, with local anchor institutions as core customers, seemed to offer a viable local opportunity.

The requirement for a review of UK agricultural policy following the UK's exit from the EU represents a significant opportunity, as well as a major challenge, to systematically reconsider the development of local food systems. Although the initial indications from the UK Government are not promising in this regard, if undertaken with a renewed emphasis on the localisation of the food system, this review could have a significant impact. The Llandovery study found that currently most of the farming around the town is upland-based, with 30 percent of produce exported −95 percent to the EU. In the post-Brexit era, this could have significant implications for the future viability of farming in the area.

Energy and energy efficiency

Over recent years, UK energy policy appears to have moved away from energy conservation and renewable generation. Feed-in tariffs for smaller energy production are reducing and there appears to be a general trend by the UK Government to look towards nuclear, imported gas, and 'fracking' as a foundation for UK energy supply. The current grid-dominated structure of UK energy production and supply makes change difficult, but the Deep Place studies have been keen to point to examples drawn from Europe, such as the Energiewende model in Germany, that show change is possible. Green energy production more readily lends itself to a localised model than the current grid-distributed fossil fuel energy production. Although there are some difficulties in bringing forward such schemes, the potential benefits from closer local and community involvement in energy production are significant. Community energy generation offers a means for communities, particularly rural communities as highlighted in the Llandovery study, to generate income streams that can be used to subsidise energy costs or other community benefits (Powell et al., 2018).

In Wales, the Environment Wales Act (2016) contains three natural resources national priorities (reiterated in the Welsh Government's *Draft Climate Change Adaption Delivery Plan*, 2018): increasing renewable energy and resource efficiency; delivering nature-based solutions; and taking a place-based approach. If successfully implemented, these priorities appear to accord closely with the approach advocated in each of the Deep Place studies and offer a fertile ground for the energy transition. Concerns were highlighted in the Llandovery study, however, about some of the difficulties in bringing community renewable schemes forward in Wales. Community renewable energy schemes represent less than 0.5 percent of Welsh renewables capacity (Jones, 2018), and this demonstrates the limited success so far achieved in developing this model of energy production in Wales.

Any move towards a more sustainable energy system must also clearly include a resolute effort to conserve energy. Energy efficiency at a community level can be significantly improved through the retrofitting of existing housing stock, but as the Tredegar study acknowledged the traditionally constructed houses of the South Wales Valleys have very poor levels of insulation and are difficult to retrofit to achieve modern levels of energy performance. Nevertheless, energy use in homes accounts for around 14 percent of UK greenhouse gas emissions, and the Committee on Climate Change (2019) has stated that the UK will miss its emissions reduction targets without near-complete decarbonisation of housing.

The Tredegar and Pontypool studies discussed the Welsh Government's Nest home energy efficiency scheme. This provides fully funded improvement measures targeted at low-income owner-occupied and privately rented households with low energy ratings. Retrofitting existing housing to improve energy efficiency would therefore seem to bring both social, and environmental benefits. In 2016, 23 percent of all households in Wales experienced fuel poverty (those that had to spend more than 10 percent of household income on household fuel costs), with 3 percent experiencing severe fuel poverty (spending more than 20 percent on fuel costs) (Beaumont et al., 2016). There is a strong consensus that living in a cold home will have lasting impacts on health, educational attainment, and the social aspects of the resident's lives. Households with older people, children, disabled people, and people with a limiting long-term illness, including people with respiratory or circulatory disease or people with a mental health condition, are most vulnerable and tend to have lower incomes (Bridgeman et al., 2016).

Energy efficiency also appears to have significant potential to contribute to local economic development. The Llandovery study highlighted a report by economist Calvin Jones (2018), which argued that, unlike renewable generation where most of the spending leaks from local areas, much of the spending on domestic re-furbishment is more successfully captured within Wales. In part, Jones suggested, this would come as a result of the more labour-intensive nature of refurbishment work, but also due to generic technologies involved that Wales has the commercial capacity to produce. The report calculated that an investment of £1.16bn in domestic refurbishment for energy saving could create 33,000 person years of employment in Wales. Over a 15-year period, this would equate to 2,200 full-time jobs on average, in areas such as construction, related skilled trades, and professional services such as surveying, planning, and engineering.

Social care

As discussed above, the communities studied have, to varying degrees, been characterised by poor levels of public health. Consequently, these areas have particularly high community care needs. The Tredegar study, which found that over 18 percent of residents were over the age of 65, identified that in 2012–2013 expenditure on social services for the county of Blaenau Gwent amounted to 26 percent of the total county council budget. This is not uncommon. Despite the scale of this expenditure, there is a widespread belief that care services, especially for the elderly, are inadequate to meet the needs of the ageing population and generally do not provide the level of service expected by older people themselves and their relatives. Nevertheless, it appears that either those with care needs or their families are reluctant to take more direct control, and the Pontypool study found that there has been only limited take-up of direct payment options.

Although local social services teams undertake most of the initial assessments for care, over 70 percent of adult community care is provided by the independent sector in the UK (UKHCA, 2013). Large care firms tend to suggest that the sector is undermined by insufficient funding provided by local authorities, which, they argue, is the cause of high staff turnover, poor working conditions, and consequential issues surrounding the quality of care provided. Burns et al. (2016) demonstrated, however, that the causes of sectoral weaknesses and the apparent unsustainability of adult community care arise not from insufficient funding, but from the profit extracted by the large corporations that generally control the sector.

In Wales, although the dominance of larger companies has been less pronounced than in England, there are nevertheless weaknesses in current provision. In Torfaen, most social care is provided by 18 different outsourced providers, and at the time of the Pontypool study, these providers were reporting that they were finding it difficult to fill vacancies. The Llandovery study highlighted a move by Carmarthenshire County Council to take over the direct delivery of some care provisions following the collapse of one of its private-sector contractors. This was an important decision by the authority, and it demonstrated its responsibility to ensure that care is ultimately delivered. The Council may ultimately decide to continue to deliver care directly as a core public service, but there are other options for it to consider in the longer term.

Each of the studies has highlighted the international experience, such as that of Bologna, concerning the benefits that can accrue from cooperative models of community care provision. The New Economics Foundation (NEF, 2013) has also shown the potential of social business models in mental health care provision, including the intrinsic value for individuals, the increased capacity and impact of public services, and the monetary value to individuals and the state. Nevertheless, despite the objectives contained within the Social Services and Well-Being (Wales) Act 2014 to create a more mixed economy of the provision of health and social care using social business models, there remain very few social business care providers in Wales (Millar et al., 2016), and this finding was certainly consistent with each of the Deep Place studies.

Moving forward

Each of the four UK Deep Place studies has been action-focused and they therefore contain a set of specific recommendations for each community. After Tredegar, common to each study is the idea of the 'coalition for change' governance model. As discussed in Chapter 7, we suggest this should become the locus for future action involving the various public, private and third sector organisations in partnership with the local community. To date, whereas each of the other locations has taken forward some of the specific action points encouraged in each study, only the Lansbury Park study has successfully prompted the creation of such a governance model. Generally, the response to each of the Deep Place studies has been more piecemeal. Although the model is still developing, the Lansbury Park experience shows that where there is effective local leadership and support from a wide range of organisations, positive change can be facilitated via this model.

The recommendations contained in each of the studies have covered a broad range of areas, including the Foundational Economy; employability strategies; personal and household debt; town centres; the future of local retail; local enterprise development; local food production; low carbon retrofitting of housing stock; distributed sustainable energy production; cooperative delivery of community care; local public sector procurement; public service delivery; community health champions; tackling crime; further and vocational education; community-focused schools; NEETs (young people not in education, employment or training); community play facilities; cultural activities; new affordable housing provision; housing allocations policy; housing evictions policy; public transport; and spatial planning. Whereas many of these are place-specific, some are common to all areas.

It is not the intention here to review the implementation of each study and every individual action point. Some actions have been taken forward in individual communities as part of other developments. Some broader policy developments have taken place across Wales that have encompassed the kind of change envisaged. Since the completion of the Tredegar study, for example, Wales' wellbeing legislation that encompasses many of the aspirations of Deep Place has been implemented and, more recently, the Welsh Government has been experimenting with the Foundational Economy that has been advocated in each study. There has also been a greater emphasis on local public sector procurement and the role of anchor institutions, and Wales has recently declared nature and environmental emergencies. To our regret, however, as we have highlighted, Wales' central economic policy focus remains on growth, and Welsh Government no longer has a target to eradicate poverty.

In many instances, the four Deep Place studies have prompted further research agendas. We hope the lasting legacy of these UK studies will be their overall impact on focusing the minds of policy-makers, practitioners, and communities, not only on the challenges and opportunities rooted in particular communities but on the broader question of what kind of society and economy we need to create to achieve economic, social, environmental and cultural sustainability. To that extent, we believe we have made some degree of impact in each of the four communities we have engaged with and in Wales more generally. Deep Place offers a practical whole place planning tool that is future-focused. Each of the studies is designed to act as a starting point, offering initial recommendations and a potential place-based co-productive governance model, but ultimately success will depend on a willingness and desire for agencies and communities to engage and commit to long-term change.

References

Adamson D (1996) *Living on the Edge: Poverty and Deprivation in Wales.* Llandysul: Gomer Press.

Adamson D and Lang M (2014) *Towards a New Settlement: A Deep Place Approach to Equitable and Sustainable Places.* Merthyr Tydfil: CREW.

Adamson D and Lang M (2017) *Lansbury Park: A Deep Place Plan.* Cardiff: Deep Place Centre.

Barry M (2011) *A Metro for Wales' Capital City Region: Connecting Cardiff, Newport and the Valleys.* Cardiff: Cardiff Business Partnership and IWA.

Beaumont A, Hulme J, Simpson E and Nowak T (2016) *The Production of Estimated Levels of Fuel Poverty in Wales: 2012–2016.* Cardiff: BRE/Welsh Government.

Bridgeman T, Thumim J, Asher M, Hodges N, Searby G and Morris P (2016) *Understanding the Characteristics of Low Income Households Most at Risk from Living in Cold Homes: Final Report to the Welsh Government.* Cardiff: Welsh Government.

Burns D, Cowie L, Earle J, Folkman P, Froud J, Hyde P, Johal S, Jones I R, Killett A and Williams K (2016) *Where Does the Money Go? Financialised Chains and the Crisis in Residential Care.* Manchester: CRESC.

Committee on Climate Change (2019) *UK Housing: Fit for the Future?* London: Committee on Climate Change.

Davies J (2007) *A History of Wales.* London: Penguin.

DfT (2011) *Personal Travel Factsheet – Commuting and Business Travel.* London: HM Government.

Egan D (2013a) *In Place of Poverty: Education and Poverty in a Welsh Community.* Merthyr Tydfil: Bevan Foundation.

Egan D (2013b) *Poverty and Low Educational Achievement in Wales: Student, Family and Community Interventions.* York: JRF.

Fabian Society (2009) *The Local Health Service.* London: Fabian Society.

IPPR North (2018) *Prosperity and Justice: A Plan for the New Economy*. Manchester: IPPR North Commission on Economic Justice.

IWA (2021) *Is This a Climate Emergency Budget?* Accessed 2 July 2021: https://www.iwa. wales/agenda/2020/01/is-this-a-climate-emergency-budget/

Jones C (2018) *The Economic Impact of Energy Transition in Wales: A Renewable Energy System Vision for Swansea Bay City Region*. Cardiff: IWA.

Lang M (2016) *All Around Us: The Pontypool Deep Place Study*. Cardiff: PLACE, Cardiff University.

Lang M (2019) *The Llandovery Deep Place Study: A Pathway for Future Generations*. Cardiff: PLACE, Cardiff University.

Lang M and Marsden T (2021) 'Territorialising sustainability: de-coupling and the foundational economy in Wales', *Territory Politics and Governance*. Doi: 10.1080/21622671. 2021.1941230

Marmot M (2010) *Fair Society, Healthy Lives: Strategic Review of Health Inequalities in England Post-2010*. London: The Marmot Review.

McClelland J (2012) *Maximising the Impact of Welsh Procurement Policy: Full Report*. Cardiff: Welsh Government.

Millar R, Hall K and Miller R (2016) *Increasing the Role of Social Business Models in Health and Social Care: An Evidence Review*. Cardiff: Cardiff.

Munday M, Roberts A and Song M (2018) *Economic Intelligence Wales Quarterly Report*. Cardiff: Welsh Economy Research Unit, Cardiff Business School.

Nathaniel Litchfield and Partners (2015) *Carmarthenshire Retail Study (2015 Update)*. Carmarthen: Carmarthenshire County Council.

New Economics Foundation (2013) *Co-production in Mental Health: A Literature Review*. London: NEF.

ONS (2020) *Regional gross disposable household income, UK: 1997 to 2018*. Accessed 1 July 2021: https://www.ons.gov.uk/economy/regionalaccounts/grossdisposablehouse holdincome/bulletins/regionalgrossdisposablehouseholdincomegdhi/1997to2018

Powell E (1884- Republished 2008) *History of Tredegar*. Blaenau Gwent Heritage Forum.

Powell J, Keech D and Reed M (2018) *What Works in Tackling Rural Poverty: An Evidence Review of Interventions to Address Fuel Poverty*. Cardiff: Wales Centre for Public Policy.

Public Health Wales and Welsh Government (2016) *Measuring the Health and Well-Being of a Nation: Public Health Outcomes Framework for Wales*. Cardiff: Welsh Government.

Sharples J, Slavin R, Chambers B and Sharp C (2010) *Effective Classroom Strategies for Closing the Gap in Educational Achievement for Children and Young People Living in Poverty, Including White, Working Class Boys: Technical Report*. London: Centre for Excellence and Outcomes in Children and Young People's Services.

Tudor Hart J (1971) 'The Inverse Care Law', *The Lancet*, 297: 405–412.

UKHCA (2013) *An Overview of the UK Domiciliary Care Sector: Summary Paper*. Wallington: United Kindom Home Care Association.

Welsh Government (2011) *Fairer Health Outcomes for All: Reducing Inequalities in Health Strategic Action Plan*. Cardiff: Welsh Government.

Welsh Government (2018) *Draft Climate Change Adaption Delivery Plan*. Cardiff: Welsh Government.

Welsh Government (August 2018) *Key Economic Statistics*. Cardiff: Welsh Government.

9

CASE STUDIES

Australia and Vanuatu

Introduction: exporting the Deep Place approach

As discussed in the previous chapter, the Deep Place approach was developed in the context of a post-industrial community in the industrial heartland of South Wales. Tredegar and similar communities such as Pontypool have a unique history grounded in the industrial revolution and the heavy industry that became the primary economic driver of the region for much of the 19th and 20th centuries. Following the Tredegar and Pontypool studies, the concept was further refined in the rural community of Llandovery. As we have seen, this provided an opportunity to develop the model and to test its practical efficacy in a rural setting. Meanwhile, the Lansbury Park study offered the possibility of completing a microanalysis of a place tightly confined by social, physical, and cultural barriers, often associated with single tenure, large council estates in the UK.

In 2016, the opportunity arose to apply the Deep Place methodology in a significantly different context of an Australian regional town. Muswellbrook is a mining community situated in the Upper Hunter region of New South Wales (NSW). The study was conducted by an Australian community housing provider, Compass Housing, which has sought to challenge the social exclusion and poverty experienced by its tenant population. In this respect, the exercise bore similarities with the study of Lansbury Park in Wales, given the commonalities of a concentration of social housing in a community characterised by low rates of economic activity and significant levels of poverty. A further opportunity arose in 2019, to apply the methodology within the significantly different context of the community of Freshwater in the Vanuatu group of Pacific Islands. Compass Housing, with which the study was again conducted, was seeking to extend a humanitarian project it had commenced in the wake of the 2015 cyclone 'Pam', which had devastated communities and infrastructure in the islands. A Deep

DOI: 10.4324/9781003221555-11

FIGURE 9.1 Map showing the location of Australian and Vanuatu case studies.

Place study was conducted to inform the priorities and actions, and to help identify the specific location to deliver the project.

The Muswellbrook and Freshwater case studies were undertaken to a more constrained set of objectives than the Tredegar and other UK case studies, which were deliberately broad in their purpose. In Muswellbrook, the overriding objective of the study was to identify employment opportunities for a specific section of the community, tenants in social housing. To this end, it ultimately proposed a wider diversification of the local economy, to create a more favourable labour market for people with limited formal qualifications and low skill levels. In particular, the study sought to identify employment opportunities within what has been described as the 'Foundational Economy' (discussed in Chapter 7), which may not necessarily eradicate local poverty entirely, but would improve welfare support payments and free many people from the bureaucratic eligibility requirements of the Australian welfare system. Increased employment, the study concluded, would also challenge local patterns of substance misuse and neighbourhood conflict, and help stabilise a highly transient population.

The Freshwater case study was also undertaken with a more targeted set of objectives than those of the UK studies. The study sought to achieve the maximum impact of a relatively limited budget for humanitarian intervention. Previous support given by Compass Housing had included the building of a community centre and cyclone refuge on Tanna Island, and the rebuilding of a community stage and rooms in Freshwater. The Deep Place methodology was deployed as an effective means to undertake an analysis of the most acute needs of the local community and to help identify where additional humanitarian interventions could achieve the most impact to improve the quality of life experienced by the community. The participatory research methods of the Deep Place study (discussed in Chapter 7) afforded a deep understanding of local culture and tradition. This ensured that proposed interventions met the expectations of local people, and supported social cohesion in a community consisting of multiple tribal affiliations and complex kinship networks.

Undertaking the studies in two very different contexts from those of the UK case studies presented a range of methodological challenges. The Muswellbrook study also applied the Atmosphere, Landscape, Horizon method (discussed in Chapter 7) utilised in Lansbury Park, but without the group of indicators identified in the UK context. Unlike the UK studies, data collection for the Muswellbrook study was often much more challenging. Although Australian census data are of an equivalent standard to that of the UK, access to non-census data in areas such as health and education was difficult. Government departments collaborated with the research to different extents, but overall, there was a general reluctance to share little more than the more widely available public data. Data provided by community housing provider, Compass Housing, therefore formed a core part of the socio-economic analysis.

The Freshwater study was completed through a mixture of remotely undertaken desk research and a week-long field visit, which facilitated direct engagement with local institutions and, most critically, enabled a community consultation. A series of face-to-face meetings were held with relevant government ministers, civil service teams, local and international NGOs, and community organisations during the field visit, as well as additional consultations undertaken by a local agent following the completion of the field work. Although the Muswellbrook and Freshwater Deep Place studies were undertaken within these more targeted objectives, much like the previous UK studies, each also reflects an overriding social ethos to positively develop people, place, and planet, and to enhance the sustainability of each settlement, a motivation that is shared by Compass Housing. Each also demonstrated a commitment to advocate, promote and deliver principles akin to those contained within the United Nations' Sustainable Development Goals and the Welsh Well-being of Future Generations Act. In this chapter, we first outline the findings of the Muswellbrook case study, before presenting those of the Freshwater study.

Muswellbrook

National and regional context

Muswellbrook is a Shire authority in the Upper Hunter region of NSW. NSW is the most populous of the Australian states with a population of 7.7 million, the majority of whom live in coastal towns and cities, with the remainder of the state characterised as rural, remote, and sparsely populated. In this context, regional towns are economically and culturally significant. As a regional town, Muswellbrook is the administrative and service centre for the Shire and is located 125 kilometres (a 1-hour and 40-minute drive) from the city of Newcastle. Newcastle is the largest non-capital city in Australia with a population of 463,000 (ABS, 2016). The city was in the past characterised by coal mining, steel production, and its port. With the demise of the steel industry in the 1980s, the port remains a primary economic driver, with an increasing role for the city as an administration centre emerging over the last ten years. It is now one of the largest coal exporting ports globally, exporting 105.9 million tonnes of blended coal in 2020 (PWCS, 2020). Destination markets include China, Japan, and India. This places it at the centre of the coal product chain, connecting it economically to Muswellbrook and the general Hunter Valley coal production region. That does not translate into a role as an employment centre for the Upper Hunter Valley, however, given poor road and rail connections that effectively place it outside of a reasonable travel to work distance.

The Muswellbrook Local Government Area has a population of around 16,000, with 12,000 attributed to the Significant Urban Area category of the Australian Census (ABS, 2016). Typical of many Australian regional towns, it is located along a major highway, in this case, the A15, which connects the coast to other major regional centres and, eventually, the remote western regions of NSW. The area of Muswellbrook South was the primary focus of the study. Although a mixed tenure area, the locality has a high proportion of public housing, managed for the state department by Compass Housing. Private sector housing consists of mainly rented properties, the majority at the lowest levels of the private sector market. Like Lansbury Park, Muswellbrook South is an edge-of-town community with tightly demarcated physical and social boundaries, encircled by external roads. There are few entry and exit points, and there is a strong sense of social and physical enclosure experienced by the residents. As with similar locations in the UK, the community experiences high levels of stigmatisation. Attitudes expressed by key local politicians cast the community in completely deficit terms, and this has been accompanied by calls for the demolition of the local social housing stock and for the relocation of the community.

Muswellbrook South is almost perfectly coterminous with the Statistical Area 1 (SA1) scale of the Australian census, and this permits a clear analysis of the patterns of poverty and social exclusion that exist in the community. In the Australian Socio-economic Index for Areas (SEIFA), Muswellbrook South lies in the top percentile of Australian SA1 areas experiencing poverty. The community experiences significant crime rates, endemic drug use, and high rates of domestic violence. Deliberate arson of properties is a major challenge. The area is not favoured by prospective public housing tenants and is usually only accepted by those with the highest needs and those facing urgent and desperate requirements for accommodation. As in the UK, this 'residualisation' (Lee and Murie, 1999) of public housing creates a concentration of populations experiencing economic inactivity and high levels of social need. Whilst to some extent improved in the UK by the adoption of choice-based letting approaches, this remains an acute problem in Australia with 82 percent of allocations of community housing in 2019–2020 defined as in greatest need (AIHW, 2021). This consequently prompted one study of the Australian social housing system to define it as an 'ambulance service' (Fitzpatrick and Lawson, 2014).

In 2017, prompted by an increasing awareness of challenging social conditions and particularly the educational impact on children, a working group was convened by the NSW government to consider the challenges that existed in the Muswellbrook community. In this context, Compass Housing undertook a Deep Place study and subsequently presented the findings to the Department of Premier and Cabinet. Consequently, a collective impact panel, similar to the 'coalition for change' (discussed in Chapter 7), was established. The CREATE Change Coalition brought together representatives of the State health and education departments, police, economic development, the Shire Council, Compass Housing, and the University of Newcastle. This panel met for two years, but, whilst triggering several actions, particularly in the field of education, it generally failed to achieve significant change within the location. A primary barrier to progress was the apparent reluctance of various agencies to collaborate over data sharing or to genuinely develop a set of common objectives for the area. The panel ceased to exist with the restructuring of the NSW government in 2019.

Challenges explored

Health

Accessing detailed health data, particularly at the lower spatial level, proved very difficult to resolve. Despite enthusiastic support for the study from key personnel within the health department, data shared were already largely in the public domain. Nevertheless, consultations with health teams at the local hospital and in the central state health department were extremely helpful in defining the pattern of issues. Data held by Compass Housing identified high levels of disability, and poor physical and mental health experienced by the tenant community. These findings were confirmed by local health personnel, who understood the health profile of the community as typical of other highly disadvantaged localities in the region. These health professionals also identified major barriers to effective engagement with the local population and they recognised how poor health literacy was a major obstacle to service engagement. Such difficulties were especially acute when seeking to engage with the aboriginal community, which had a collective, historically negative experience of public services.

An unpublished 2015 New England health district report that was shared with the research team, identified obesity, smoking behaviours, fall-related injuries, and preventable admissions as major challenges for the local hospital. This was also highlighted in consultations with the local health team, who confirmed that much of their service provision addressed drug and alcohol-related admissions, domestic violence, and low birth weights. Child health was a particular concern of the local health community and this became the focus of a specific initiative, the Healthy Kids Bus Stop. This service is a rural health initiative bringing child health experts directly into the community, followed up with referrals to a relevant specialist centre or hospital. Services include audiology, oral health, dietary assessment, and occupational and speech therapy. A core concern was that in some of these fields, nearly 51 percent of referrals do not follow through for assessment and treatment. This presents lifetime health implications for the children who do not receive the treatment they need. Key barriers to attendance at referral include physical distance to the point of referral, which can often involve an overnight stay with associated accommodation, and transport costs. This is a common problem in Australia where the lack of remote and regional services is the product of low population levels.

The historical treatment of First Nations families and the cultural trauma legacy of the 'stolen generations', add a complex additional barrier that is reflected in the wider ill-health profile of aboriginal communities. This is the subject of a longstanding federal government programme called Closing the Gap (Commonwealth Government, 2020), but annual evaluation reports have generally demonstrated a failure to make significant progress (Perche, 2017). The most recent implementation plan included revised targets for child health that address low birth weight and child development, which are two of the most significant contributors to poor health outcomes for the Aboriginal and Torres Strait Islander population (Commonwealth Government, 2020). For Muswellbrook South, a knot of health challenges exists that fixes poor child and adult health as a key feature of the community. Actions included in the CREATE Change Coalition 2019 Action Plan included the introduction of a Muswellbrook Healthy and Well mental health and well-being programme, the delivery of a national mental health programme (Act Belong Commit), and an Aboriginal Chronic Disease programme. There were also intentions to improve service delivery and accessibility. Given the demise of the CREATE Coalition, however, there are no reporting frameworks to assess the impact of these interventions.

Education and skills

Poor educational attainment of children in the community became a central concern of the CREATE Change Coalition. Participation of the University of Newcastle, NSW Family Action Centre, brought clarity and focused action in this public policy domain. A high proportion of early school leavers compared with Australian norms, and low levels of adult educational qualifications were significant factors in the low rates of economic activity of social housing tenants in Muswellbrook South. This, when combined with poor mental and physical health, impacts the 'triangle of poverty' we discussed in Chapter 3. Evidence provided to the CREATE Change Coalition by NSW Family Action Centre heightened concerns about the educational achievement of children living in the community. The Index of Socio-Educational Advantage identifies factors including parental occupation and education, school remoteness, and proportion of indigenous students. According to this index, all the schools serving the Muswellbrook South community were below average. Similar findings were evident in the Australian Early Development Census, which exposed vulnerability in the domains of physical health and well-being, social competence, emotional maturity, language, and cognitive and communication skills. The proportion of children with vulnerability in one or two domains was also higher than Australian norms.

In an approach centred on long-term prevention, educational improvements featured significantly in the discussions and actions of the Coalition. The University of Newcastle committed student placements in occupational therapy to the locality and opened a Tertiary Education Centre in Muswellbrook, which sought to provide open access to learning opportunities for the community. In addition, the Education Authority developed a Complex Case Coordination programme to coordinate inter-agency interventions with children experiencing major educational challenges and a history of trauma. The University of Newcastle convened and co-managed the collective impact project with the Muswellbrook Shire Council. This project funded a member of the Council's community engagement team to act as a Collective Impact Facilitator, to deliver programme objectives of preventative interventions to address negative child health and behavioural outcomes in marginalised communities. Based on the CREATE-ing Pathways to Child Wellbeing programme developed at Griffith University (Branch et al., 2020), the Strong-Families – Capable Communities programme focused on local collaboration and service integration.

Housing

The social housing stock of Muswellbrook South was built during the 1960s and 1970s to provide accommodation for workers employed at the two coal-fired power stations in the region. In the transition to social housing, lack of investment has created a repairs backlog and, consequently, the community now experiences poor quality housing. Although the housing stock was owned by the Families and Communities Services department (now called Department of Communities and Justice) of the NSW government, at the time of the study it was managed by Compass Housing on a three-year 'fee for service' contract. This arrangement frustrated housing investment and the development of a long-term maintenance programme. In 2019, however, a management transfer to a 20-year contract improved the ability of the housing provider to invest in repair. Unfortunately, without formal title transfer, the ability to invest continues to be limited in comparison to stock transfer experiences in the UK (see discussion in Chapter 8). The average per dwelling spend on repairs at the time of the study was AUD4,818 per annum, compared with an average of the remainder of housing stock managed by

Compass Housing of AUD2,518. This helps illustrates the scale of the long-term repair backlog, but also reflects a high level of tenant damage.

Data held by Compass Housing in relation to its tenanted properties has provided additional evidence of the level of poverty that was indicated by Census and SEIFA data. The data showed that just 3.2 percent of tenants derived their primary income from salaries or wages. Nearly 31 percent of tenants were in rental debt, with an average debt of AUD340, whilst nearly 40 percent of tenants also experienced water utility debt. Overall, the tenant population was defined by low-income groups. The largest single category of tenants were single-person households (33 percent) (predominantly male) and sole parents (41 percent) (predominantly female). Some 77 percent of tenants reported a personal or family disability. Over 43 percent of tenants were allocated a property from a situation of homelessness or risk of homelessness, and 57 percent were designated in 'Greatest Need' in the allocations system. Whilst many of the patterns of social disadvantage found in Muswellbrook South were comparable with those experienced in several of the UK Deep Place studies, there were, however, unique characteristics identified in Muswellbrook South. One of the most apparent differences was the levels of disadvantage experienced by those residents who identified as of Aboriginal or Torres Strait Islander origin. These constituted 26 percent of tenants, which compared with an 8 percent proportion in the wider local community and a national proportion of 3.3 percent (AIHW, 2021).

The physical location of housing stock compounded these patterns of economic poverty and created complex patterns of social exclusion. Situated on the town's periphery, 1.8 kilometres distant from public services located in the town, there were, unsurprisingly, low levels of community service engagement by residents of Muswellbrook South. Moreover, there was considerable dissatisfaction in the community with the quality of public services available. Issues of particular concern that were identified by the study's Landscape Assessment included: poor street lighting, a 'play' area where all equipment had been removed, and major delays removing fly tipping or accumulations of non-routine domestic waste. Community Think Spaces confirmed the presence of high levels of substance misuse, domestic violence, and neighbourhood conflict, which had been identified by the data analysis. Participants in these community sessions were themselves directly living with addiction, family breakdown, and deep poverty. Older members, who formed the core of a stable community, reported harassment, high fear of crime, and considerable personal insecurity. These experiences were confirmed by the local police command which, although unable to provide crime data at a relevant spatial level, confirmed that Muswellbrook South was indeed considered an area experiencing high rates of crime and anti-social behaviour.

The favoured strategy of the local housing provider to help take forward the findings of the Deep Place study was to convert a residential property into a 'community hub', a strategy it had employed successfully in other similarly disadvantaged locations. Despite two years of negotiation with the state department, however, it failed to overcome resistance and acquire the necessary change of use planning approval. Consequently, the community hub strategy was abandoned in favour of a three-year area-based community development strategy. This strategy has so far led to intensive interventions in several housing complexes that were experiencing significant levels of anti-social behaviour, and these have, to date, been successful in reducing neighbour disputes. The provision of social facilities, such as barbecue stands in common areas, has improved sociability and social cohesion, and such measures are beginning to stabilise the previously highly transient population.

Transport

As found in the UK Deep Place studies discussed in Chapter 8, limited and inequitable access to transport can reinforce patterns of poverty and social exclusion. Poor communities, particularly in more remote locations, tend to have a significantly higher dependence on public transport as their primary means of connectivity. Such communities, however, also tend to experience low frequency, inconveniently scheduled and routed, and disproportionately expensive public transport. Muswellbrook South follows this general pattern and the impact of poor access to public transport is experienced by the community as a barrier to accessing basic public services.

Although local journeys from the community to the town centre are supported by two bus routes, services are infrequent and generally only available before midday. Residents reported a severe lack of bus services at certain points in the day, particularly during the evening, which prevented them from accessing leisure facilities in the town centre. Think Spaces held during the research also identified the limited public transport provision to enable residents to access medical and other support appointments. With a distance of 2.4 kilometres to the town centre, walking was difficult, especially during summer months when temperatures are routinely over 30° C. Moreover, it was reported anecdotally that some bus drivers refused entry to indigenous people, although this perception could not be verified many community members felt it to be the case.

Longer journeys between Muswellbrook and Newcastle are generally undertaken by train, a journey of around two hours. Given the comparatively short physical distance between the two settlements, this journey duration is symptomatic of the lack of rail investment in the region, which is characterised by an older and slower rolling stock and track system. A lack of integration with local bus services further reduces the viability of a commute to employment in Newcastle from residents of Muswellbrook South. The Deep Place study thus concluded that a lack of access to affordable and efficient public transport was a barrier to employment, access to public services, and wider social engagement for the residents of Muswellbrook South, which often added to the social isolation and stigma experienced by the residents.

Sustainable economic activity

At the time the Muswellbrook South study was undertaken in 2016, mining was by far the most significant local employer, with over 28 percent of regional employment located within the sector. Whilst this employment pattern is to some extent cyclical, rising and falling with the global demand for coal, there is a strong underlying mining culture and identity, much like that historically found in the South Wales Valleys. The industry is now predominantly based on strip mining on a vast scale, and the Muswellbrook South community sees its future economic viability intricately connected to the survival of the industry. The decline of employment in the sector is inevitable, however, as it faces existing pressures of automation and the decline in global coal demand that will be felt over the next decade. The latter is likely to result from the COP26 agreement of 23 nations to 'phase down' coal use (UK COP 26, 2021). This interpretation was quickly refuted by the Australian Prime Minister within days of the COP 26 agreement, who promised continued exploitation of Australian coal reserves (Clarke, 2021).

Despite the political gesturing of the Australian Prime Minister, however, the inevitable transition away from coal mining resulting from a global reduction in coal demand has been recognised by several of the trade unions representing the industry. Since the Deep Place study was undertaken, a coalition of seven trade unions and several regional community organisations has formed the Hunter Jobs Alliance, which aims to pursue economic diversification and foster a wider understanding of the need to transition from an economic dependency on coal (HJA, 2021). Nevertheless, the concept of transition remains politically sensitive and has become a touchstone of a long-running 'climate war' between the major political parties, causing one commentator to conclude: 'in my view, progress is hampered by some at one end of the debate who want no change at all, and others at the other end who want to move faster than is feasible' (Finkel, 2021). Australian politics has consequently become deeply divided over the issue of coal mining, with metropolitan areas far more sympathetic to climate change mitigation and phasing out of fossil fuels. Despite these tensions, an important objective of the Muswellbrook South study was to propose innovative patterns of economic activity that could support a transition away from the coal-dominated economy of the region.

Food production and supply

At the time of the study agriculture employed 5.4 percent of the working population. Viticulture, for which the region has a global market, accounted for a large proportion of this employment. As well as being a major export industry, the presence of the 'cellar door' food and wine tasting venues contribute more generally to the visitor economy that is based on a regional reputation for fine food experiences. The Deep Place study proposed building on this reputation to develop high-value product chains based on charcuterie products and organic foods. Free-range poultry and egg production were identified as having the potential to capitalise on increasing market demand for organic products that have been predicted by the Australian Organic Association. The study proposed the collective branding and marketing of a Hunter Region organic trademark, with the potential to expand production and employment in organic food production. The Chinese export market for safe, reliable foods was particularly noted at the time the study was undertaken (Zonca, 2016). This market has recently diminished as a result of embargoes on the import of Australian wines and barley into China, which is symptomatic of declining political and trade relations between the two countries (Wagner and Zhou, 2020). Further opportunities were suggested during the study for several other areas, including cotton, biofuel, and plant-based pharmaceuticals. More recent concerns about the water demands of cotton production and land use losses for biofuels, however, would now preclude their inclusion in a sustainable economic strategy.

Energy and energy efficiency

Located in close proximity to Muswellbrook, the Liddell and Bayswater coal-fired power stations provide a significant contribution to the baseload requirements of NSW state and Australia's power supply. The planned closure of the Liddell power station in 2023, offers an important opportunity for energy diversification. The planned closure became a national political issue within Australia, and the federal government was reported to have pressurised energy company AGL to maintain the life of the Liddel power station beyond its planned closure date (Grattan, 2017). Recently announced plans for a gas-fired power station and hydrogen economy proposals have dissipated this political issue. The nearby Bayswater coal-fired power station is also scheduled for closure in 2035 (ABC News, 2017). The Deep Place study highlighted the opportunity presented by the closure of these power stations for the Hunter region to become an Australian exemplar for renewable energy based on wind and solar energy production. Opportunities were also identified for clean burn household waste generation technology and the associated reduction of landfill waste. Linked to the possible development of agricultural animal husbandry, bio-gas energy generation was also identified as an area for further research. Recently, the development of large-scale hydrogen energy production has been proposed for the region (Vorrath, 2021), however, the 'green hydrogen' credentials of the proposal have yet to be proven.

Health and social care

Employment in the sectors most closely associated with what has been termed the 'Foundational Economy' (Bentham et al., 2013) (see discussion in Chapter 7) accounted for over 42 percent of local employment in Muswellbrook South (ABS, 2016). Consequently, the Deep Place study proposed the development of a Foundational Economy strategy, grounded in the economic activity created by local anchor institutions, including the local authority and the hospital. The care sector was specifically identified as having potential for the targeted recruitment of social housing tenants. Expansion of the elder-care sector, childcare services, and the launch of the National Disability Insurance Scheme all projected considerable growth in the care economy. The lower qualification entry points, and the opportunities for professional training and upskilling in the sector, were identified as a potentially significant element of the proposed strategy. The study also recommended the development of care cooperatives to promote local ownership and control, as well as to help overcome the often exploitative employment conditions and low wages experienced in the for-profit sector.

The visitor economy

In exploring the potential to develop the visitor economy, not all options considered as part of the Deep Place study were considered deliverable as fully sustainable models. Building a day visitor attraction specifically for Muswellbrook, for example, required significant linkage to existing assets and activities in the wider region to become viable. Muswellbrook was not generally included in regional wine tours, motorcycle tour routes, cruise ship day visits, or the short stay domestic tourist market from Sydney. Nevertheless, local physical and natural assets do offer significant opportunities to develop further as part of a visitor's offer. Local opportunities for development exist in the wine economy, as well as developing the musical and arts identity of the town around the Regional Arts Centre and the Conservatorium of Music. Creating stronger links to regional national parks for servicing walking-based activities were also considered as part of the study. Creating a distinct identity for Muswellbrook Shire was, ultimately, seen as an important prerequisite to attracting visitors.

Vanuatu

National and regional context

The nation of Vanuatu consists of a group of islands in the Pacific region of Melanesia, which also includes Fiji, New Caledonia and the Solomon Islands. Port Vila, the capital of Vanuatu, is located 2,481 kilometres (a 3.5-hour flight) from Sydney. The country includes over 80 islands, though only the largest are inhabited. Port Vila is located on the island of Efate. At the time the Deep Place study was undertaken the primary sources of demographic data were the 2009 Census and a Mini Census conducted in 2016. The total population of Vanuatu in 2016 was a little over 270,000, over 99 percent of whom were of Melanesian descent (VNSO, 2016). Approximately 25 percent (57,000) of the population lived in urban locations with 44,000 located in the capital Port Vila, which is the primary destination for internal migration (VNSO 2009). The subsequent 2020 Census records a population growth of 30,000 in Vanuatu since the previous census (the urban population actually declined by just under 2 percent) (VNSO, 2020).

As well as being an administrative centre, Port Vila is also a major holiday destination, particularly for Australians, and is part of the Pacific cruise circuit. Although this has led to the development of the harbour area, with its restaurants and tourism infrastructure, a short journey away from the coast are the informal settlements more typical of the Pacific region. The Deep Place study (and associated humanitarian programme) was conducted in the relatively mature informal settlement of Freshwater (Fres Wota), which has a population of around 10,000, consisting of multiple tribal groups. In common with much of the Pacific region, the nation is particularly vulnerable to earthquakes, tsunami, volcanic eruptions, and tropical cyclones. In recent times, considerable devastation has been caused by cyclones 'Pam' (2015) and 'Harold' (2019). Consequently, recovery from natural disasters poses major challenges for the nation.

Challenges explored

Health

Vanuatu and the community of Freshwater face considerable health challenges, which have been described by the WHO as a 'triple burden' of non-communicable and communicable diseases, and climate change-related health threats (WHO, 2018). Much of the health burden derives from rapid urbanisation and the movement of island populations into the urban centres. Freshwater typifies this pattern, with a poor urban environment lacking basic public and civic amenities. Transmission of communicable diseases including tuberculosis, sexually transmitted diseases, dengue, and measles is more likely in the overcrowded and insanitary conditions of informal settlements such as Freshwater. Rising rates of cancer, diabetes, and respiratory disorders are particularly prevalent amongst the adult population, and diarrhoeal diseases are a major cause of child illness and mortality (WHO, 2018).

At the time that the Deep Place study was undertaken, the Vanuatu government's response to this public health landscape was encapsulated in the Health Sector Strategy (HSS, 2017–2020) (Ministry of Health, 2017). This is linked to the Society Pillar of the National Sustainable Development Plan (DSPPAC, 2016), and, particularly, to its Goal 3 to create a 'healthy population that enjoys a high quality of physical, spiritual and social well-being' (p. 3). Consultation with the Council of Chiefs during the Deep Place research identified additional needs for dental, optometry, and basic nurse-led medical services. The Deep Place study identified the continued use of the previously rebuilt community stage as a potential 'hub' offering health outreach services, but whilst the facility was furnished with the basic requirements of a consultation and treatment space in 2020, COVID-19 travel restrictions have frustrated the development of delivery partnerships to provide services.

Housing

Freshwater is a relatively mature informal settlement with communal and house-hold water and electricity supply. For Port Vila as a whole, 46 percent of homes have access to a direct water supply, whilst 23 percent have a shared water supply (VNSO and UNDP, 2013). In contrast with rural areas, where pit latrines pre-dominate, 68.2 percent of Port Vila homes have access to a flush toilet. Fresh-water is likely to experience more limited access to individual household water supply and sanitation facilities than the Port Vila data suggest, but it was not possible to quantify access at this spatial level. Moreover, housing in Freshwater is generally poor and highly vulnerable to disaster events such as tropical cyclones. In contrast to much of Vanuatu, where the majority of home construction is from traditional materials, 90 percent of housing stock in Port Vila is constructed from concrete flooring, block walls, and corrugated iron sheets (VNSO, 2016). In 2013, 45 percent of homes in Port Vila were classed as 'permanent', whilst 27 percent were 'makeshift' and 18 percent 'traditional' (VNSO and UNDP, 2013).

Whilst road access to the community is generally good, internal roads are unsurfaced and subject to flooding, potholes, and deep ruts. Overall, the general urban environment is degraded, with rubbish accumulation, multiple abandoned vehicles, and frequent small roadside fires burning to avoid rubbish collection fees. Although a single-use plastic ban has reduced the plastic fume content of such fires, the wider consequences for air quality and, consequently, health outcomes

FIGURE 9.2 A typical Freshwater street scene. Photograph by author.

were still evident at the time the Deep Place study was undertaken. The failure of a small-scale international housing trial in Freshwater, where two brick and tile constructed pilot homes failed rapidly in the tropical conditions, was noted during the study. The solution identified would involve a closer understanding of local traditional construction methods and materials that have demonstrated longer-term resilience, coupled with modern techniques to achieve Category 5 cyclone proofing.

Education and skills

In many of the island communities of Vanuatu, access to educational facilities is severely limited, and even in the capital, Port Vila, completion of the tertiary-level education is just 6.6 percent. Particularly notable are the dropout rates between primary school to junior secondary, and later in the transition to senior secondary education. Of those taking up primary education, 61 percent will not enter junior secondary school, and a further 20 percent leave at the senior high school transition point (VNOS, 2020). Self-reported literacy rates are high, with 82 percent of the population reporting literacy in Bislama, 67 percent in English, and 37 percent in French. Nevertheless, research undertaken by the World Bank (2014), noted that one community with a self-reported literacy rate of 85 percent in reality had a rate of just 27 percent.

In general, poverty and the inability to pay school fees and transport costs, as well as boarding fees for residents from remote island locations, are significant causal factors in the low education participation rate in Vanuatu. The Deep Place study was unable to address these wider structural conditions, but it did conclude that the development of the Freshwater Community Stage as a community hub may enable informal learning opportunities for that particular community and, if this is successful, it may serve as a model elsewhere in Vanuatu. Moreover, the research team noted the work of Wan Smol Bag, which is a Vanuatu youth initiative delivering learning and participation opportunities to young people in Port Vila. It is hoped that an outreach service as part of this project might be delivered at the Freshwater community stage hub.

Gender

The traditional Pacific social structures afford considerable authority to men, and in Freshwater, this was evident in the composition and role of the Council of Chiefs. The Council consisted of chiefs from up to ten tribal communities residing in the Freshwater community. The Council's role is central to the maintenance of social cohesion and to addressing social issues, and the traditional authority of the chiefs remains intact despite cultural changes associated with rapid urbanisation. A gang culture had emerged around 2010, for example, which was identified as the cause of rising crime rates and drug use at the time. Interventions by the chiefs, it was claimed, significantly reduced the problem and the crime rates returned to a more conventional level. One of the younger chiefs contributing to the consultation activities of the Deep Place study, who had himself been a member of the 'Vietnam gang', largely credited the interventions of chiefs as undermining the developing gang culture.

Less clear from these traditional social structures is the support given to the many problems experienced by women. The Port Vila Women's Centre (VWC and VNSO, 2011) has highlighted the prevalence of domestic and family violence, rooted in wider gender inequality in the community. The Women's Centre regarded widespread and commonly held beliefs about gender roles as undermining women's human rights. It saw the traditional management of sexual violence as inadequate, rarely resulting in legal redress and often returning perpetrators of family violence to the family home. Furthermore, the Women's Centre regarded violence against women as normalised within the community, and therefore a source of considerable under-reporting of the experiences of women in Vanuatu society. The incidence of violence against women in Vanuatu is reported in Table 9.1. Whilst preserving the necessary social cohesion function of the Council of Chiefs was seen as central to local cultures, the Deep Place study also saw the urgent need to strengthen the position of women and improve their security from violence. This need was supported as a major part of the development project through market accommodation for women referred to in the following section.

Sustainable economic activity

Recent economic activity data for Vanuatu was difficult to identify and, consequently, the Deep Place study was heavily reliant on the 2009 Census, the 2016 Mini Census, and a 2014 World Bank report (itself reliant on the 2009 Census). Nevertheless, it was evident that even in urban areas, traditional subsistence food production remains a major source of economic activity in Vanuatu. The World Bank has estimated that 45 percent of the population of Vanuatu is engaged in food production for personal consumption (World Bank, 2014). The more rural islands are dominated by informal economic activity in subsistence husbandry, agriculture, and fishing. Formal employment is higher in urban areas,

TABLE 9.1 Violence against women in Vanuatu

Type of violence	Women experiencing violence during lifetime (%)
Physical or sexual violence	60
Emotional violence	68
Coercive control	69
Rural rate of incidence	63
Urban rate of incidence	50

Source: VWV and VNSO, 2011.

TABLE 9.2 Main sources of household income in Vanuatu (%)

	Wages/ salary	Land lease	Remittances	House rent	Sale of fish, crops, handicrafts	Own business	Other
Vanuatu	28	0.2	7.1	1.3	41.6	16.1	4.8
Urban	64.4	0.1	3.8	4	10.2	15.5	1.5
Rural	16.7	0.2	8.1	0.4	51.4	16.3	5.8
Port Vila	66.6	0.1	3.9	4.3	10.1	13.7	1

Source: Post Pam Mini Census Report, Volume 1, 2016.

with Port Vila recording a 62.8 percent labour participation rate in the 2009 Census and high levels of wage labour as the primary income source in the 2016 Mini Census. Remittances from migrant labour are an important element of family incomes, particularly in rural communities. Largely young male workers participate in migrant worker schemes in Australia and New Zealand, returning income to their families. Although largely seasonal patterns of employment, these make a critical contribution to the local economy and have been important sources of work during the COVID-19 pandemic despite evidence of more extreme exploitation than evident pre-pandemic (Stead, 2021).

In Freshwater, the local market has played a pivotal role in both the local economy and food supply. Supermarkets in the town centre were almost exclusively used by more affluent members of the community, or by the personnel of the many agencies and NGO offices in the area. The residents of Freshwater almost exclusively secured food supply from their own small-scale production and from the local market. The function of such markets is common in the Pacific region, and they can offer important opportunities for women to become economically independent (UN Women, 2021). Although covered, but open sided, the Freshwater market has limited facilities. It is, however, preferred by female traders as it is less competitive and crowded than the main town centre market. The market is centrally located within the Freshwater community at an intersection of three local roads, and around 400 metres from the community stage building.

A female market manager works with the female vendors and there is a collaborative and convivial atmosphere. The important issue for the vendors is that, given the distances they travel to bring products to market, it is often necessary for them to stay several nights until their produce is sold. Frequently accompanied by children, traders sleep under the market cover but are effectively in the open air. There are many risks including extreme weather and flooding, but women traders are at real risk of sexual violence and theft of produce. The Deep Place study concluded that the marketplace was a core component of the Foundational Economy in Freshwater and could provide a platform for further economic development. The provision of secure accommodation for the women traders, therefore, would make a significant contribution to their safety and to the functioning of the market as a local economic hub. A design has subsequently been commissioned,

FIGURE 9.3 Freshwater market. Photograph by author.

which when complete will accommodate up to eight women and four children, with toilet, shower, and limited cooking facilities. Most importantly, the security of the building will be controlled by the women traders themselves. This supports both social and economic independence for the women and helps to balance the traditional gender structures of the community. The implementation of much of the Freshwater programme envisaged in the Deep Place study has been badly delayed by COVID-19-related travel restrictions. It will be resumed as soon as conditions permit, and subsequent delivery will be identified through further field visits.

Moving forward

Although undertaken in substantially different contexts from the UK, the Muswellbrook and Freshwater studies have demonstrated that Deep Place can be a valuable and effective research and policy tool in a wide variety of settings. The most important conclusion that may be drawn from the Australian and Vanuatu experience has been that effective research is heavily reliant on access to recent data at an appropriate spatial scale. The Muswellbrook and Freshwater studies have encountered significant challenges in meeting this requirement. In Australia, basic publicly available data sources include a five-yearly census and various standard large-scale household and employment data sets, which mirror those in the UK. There is, however, a considerable reluctance by public sector organisations to share their own data, and the failure of the collective impact programme in Muswellbrook can be largely attributed to this cultural reluctance to cooperate. As we have discussed, limited data availability, particularly at the spatial level of the Freshwater community, was a major obstacle in the Vanuatu study. This highlighted the need for effective local consultation at the government and community levels, which the Deep Place approach was helpful in redressing. The approach proved informative in providing an understanding of the patterns of community life and community priorities in Freshwater. The Deep Place method proved to be adaptable to very different, socio-economic and cultural contexts. It worked effectively at different spatial scales, as, at its centre, lies a commitment to building sustainable and equitable places.

References

ABC News (2017) *Liddell Power Station: AGL conforms closure of coal plant, replaces it with renewable energy.* (December). Accessed 29 October 2021: https://www.abc.net.au/news/2017-12-09/liddell-coal-plant-closes/9243180.

Australian Bureau of Statistics (ABS) (2016) *2016 Census.* Accessed 12 November 2021. https://www.abs.gov.au/websitedbs/censushome.nsf/home/2016.

Australian Institute of Health and Social Welfare (AIHW) (2016) *Snapshot. Profile of indigenous Australians.* Accessed 9 December 2021: https://www.aihw.gov.au/reports/australias-welfare/profile-of-indigenous-australians.

Australian Institute of Health and Social Welfare (AIHW) (2021) *Web report: Housing assistance in Australia. June 2021.* Accessed 3 December 2021: https://www.aihw.gov.au/reports/housing-assistance/housing-assistance-in-australia/contents/priority-groups-and-waiting-lists.

Bentham J, Bowman A, de la Cuesta M, Engelen E, Ertürk I, Folkman P, Froud J, Johal S, Law J, Leaver A, Moran M and Williams K (2013) *Manifesto for the Foundational Economy.* Manchester: Centre for research on Socio-Cultural Change, University of Manchester.

Branch S, Homel R and Freiberg K (2020) *CREATE-ing Pathways to Child Wellbeing in Disadvantaged Communities – Phases 2 and 3 (2016–2020): The Collective Impact Facilitator Role – Overview.* Brisbane: Griffith Criminology Institute. Accessed 23 November 2021: https://www.griffith.edu.au/__data/assets/pdf_file/0029/1179452/CIF-Report-Phase-2-3-Overview-FINAL-06102020.pdf#:~:text=The%20CREATE-ing%20Pathways%20to%20Prevention%20Initiative%20%28CREATE%29%20was, behaviour%2C%20substance%20misuse%2C%20and%20involvement%20in%20youth%20crime.

Clarke M (2021) 'COP26 agreement to phase down coal not a 'death knell' for coal power says PM, disputing Boris Johnson', *ABC News.* Accessed 20 November 2021: https://www.msn.com/en-au/money/markets/cop26-agreement-to-phase-down-coal-not-a-death-knell-for-coal-power-says-pm-disputing-boris-johnson/ar-AAQHOxN.

Commonwealth Government (2020) *National Agreement on Closing the Gap.* Accessed 26 November 2021: https://www.closingthegap.gov.au/national-agreement/national-agreement-closing-the-gap.

Department of Strategy, Policy and Planning and Aid Coordination (DSPPAC) (2016) *Vanuatu 2030, The Peoples Vision. National Sustainable Development Plan 2016–2030.*

Finkel A (2021) 'Getting to zero: Australia's energy transition', pp. 6–161 in: C Feik (ed.) *The Quarterly Essay*. Carlton VIC: Black Inc Books.

Fitzpatrick S and Pawson H (2014) 'Ending security of tenure for social renters: transitioning to "Ambulance Service" social housing', *Housing Studies*, 29(5): 597–615.

Grattan M (2017) 'Government leans on AGL over Liddell ahead of meeting', *The Conversation*. Accessed 3 March 2018: https://theconversation.com/government-leans-on-agl-over-liddell-ahead-of-meeting-83778.

Hunter Jobs Alliance (HJA) (2021) *Building for the future: A 'Hunter Valley Authority' to secure our region's prosperity*. Accessed 29 August 2021: https://static1.squarespace.com/static/5f9b9768d62e163b28e5edf5/t/60d2e2755b043e4141b1b078/1624433272281/HJA+2021_Building+for+the+Future_A+Hunter+Valley+Authority_lores.pdf.

Lee P and Murie A (1999) 'Spatial and social divisions within British cities: beyond residualisation'. *Housing Studies*, 14(5): 625–640.

Ministry of Health (2017) *Heath Sector Strategy (HSS) 2017–2020*. Port Vila: Government of Vanuatu. Accessed 17 December 2021: https://www.dfat.gov.au/sites/default/files/-vanuatu-health-sector-strategy-2017-2020.pdf.

Perche D (2017) 'Closing the Gap is failing and needs a radical overhaul', *The Conversation*. Accessed 1 December 2021: https://theconversation.com/closing-the-gap-is-failing-and-needs-a-radical-overhaul-72961.

Port Waratah Coals Services (PWCS) (2020) *Financial Report 2020*. Accessed 3 December 2021: https://pwcs.com.au/media/2514/financial-report-2020.pdf.

Stead V (2021) 'Australia needs better conditions, not shaming, for Pacific farm workers', *The Conversation*. Accessed 28 February 2022: https://theconversation.com/australia-needs-better-conditions-not-shaming-for-pacific-farm-workers-171404

UK Cop26 Presidency (2021) 'End of Coal in Sight at COP26', *United National Climate Change press release on behalf of the COP26 Presidency and the COP26 High Level Climate Champions*. Accessed 3 December 2021: https://unfccc.int/news/end-of-coal-in-sight-at-cop26.

United Nations Women (2021) *A Toolkit for Operating Markets*. New York: United Nations.

Vanuatu National Statistics Office (VNSO) and the United Nations Development Program (UNDP) (2013) *Vanuatu Hardship and Poverty Report. Analysis of the 2010 Household Income Survey*. Suva, Fiji: UNDP Pacific Centre.

Vanuatu National Statistics Office (VNSO) (2016) *Post-TC Pam Mini Census Report Volume One*. Port Vila: Ministry of Finance and Economic Management.

Vanuatu Women's Centre (VWC) and Vanuatu National Statistics Office (VNSO) (2011) *Vanuatu National Survey on Women's Lives and Family Relationships*. Port Vila: VWC.

Vorrath S (2021) 'Hydrogen Valley: Plan unveiled to turn Hunter into a renewables hydrogen hub', *Renew Economy.com.au*. Accessed 3 December 2021: https://reneweconomy.com.au/hydrogen-valley-plan-unveiled-to-turn-hunter-into-a-green-hydrogen-hub/.

Wagner M and Zhou W (2020) 'It's hard to tell why China is targeting Australian wine. There are two possibilities', *The Conversation*. Accessed 14 November 2021: https://theconversation.com/its-hard-to-tell-why-china-is-targeting-australian-wine-there-are-two-possibilities-144734.

World Bank (2014) *Vanuatu: Socio-Economic Atlas*. Washington, DC: World Bank. https://openknowledge.worldbank.org/handle/10986/18669. Accessed 12 December 2019.

World Health Organisation (WHO) (2018) *Country cooperation strategy: At a glance*. Accessed 17 December 2021: http://apps.who.int/iris/handle/10665/136982.

Zonca C (2016) 'Selling the story of provenance key to Aussie farmers cracking the Chinese Market', *ABC News*. Accessed 22 January 2017: https://www.abc.net.au/news/rural/2016-04-14/provenance-key-to-cracking-china/7327042.

10
FINDING A REGENERATIVE SOCIAL, ECONOMIC, AND POLITICAL ORDER

Introduction

The relationships between political systems, economic paradigms, and social structures are complex and historically contingent within and between nations. The primary representation of a struggle between capitalism and socialism is one that has conditioned our historical understanding of nation-specific and global politics. In this context, the hegemony of the neoliberal paradigm, both as an economic model and a political model, is extreme. The celebration of the economic market, as the primary means to allocate public and private goods and rewards, has its parallel in a political system that recognises the individual consumer as a sovereign being, capable of making rational choices within the framework of the market. Communities and, indeed, families are relegated to an agglomeration of individual choices rather than the expression of any communal will. Whilst Fukuyama's (1992) claim for the 'triumph of capitalism' over all alternatives was premature, exaggerated and itself an outcropping of the neoliberal ideology, the ability of democratic socialism to articulate an alternative has been severely curtailed. In this sense, the traditional models of socialism and communism have been rendered historically specific, and it is difficult to see their re-emergence in recognisable forms as any clear force for change in the quest for social and climate justice. We would also add that the multiple failures of actual socialist and, particularly, communist 'experiments' do more to delegitimise key outcomes of socialist thought than Fukuyama's triumphalism. From the horrors of Stalin's enforced collectivisation of agriculture to contemporary human rights abuses in China, extreme caution is required on the road to a replacement or radical reform of capitalism. For our purposes here we also recognise that socialism, and certainly Soviet and Chinese models of communism, have historically embraced the concept of rapid economic growth and carbonism as the primary

DOI: 10.4324/9781003221555-12

means of opposing capitalism in the world economic order. The disjuncture between even social democracy and environmentalism is yet to be resolved by any semblance of a red and green political paradigm

In undertaking the first of our Deep Place studies in Tredegar in Wales (discussed in Chapter 8), we called for a 'new settlement' on a scale equivalent to the post-WW2 reforms, with the capacity to eradicate poverty and achieve a zero-carbon economy. We reaffirm that call here, and, furthermore, suggest that a new global settlement is needed, one that is capable of building on the various declarations of the United Nations and making the achievement of the Paris Agreement and the Sustainable Development Goals (SDGs) a reality. This requires a challenge to the hegemony of neoliberalism and to the current model of capitalism, which seek to sidestep these crises with 'carbon offsetting' logics and marketised palliatives by the financial sector. Consequently, we must now ask: *where will the basis of a socio-political mode of organisation and action be found to challenge the current levels of global inequality and address the global environmental crisis?*

At their core, most ideologies that have confronted capitalism tend to embrace a conception of collectivism. It is the contention of this book that we need to formulate a new form of collectivism, that embraces not only our human relationships but also our relationship with nature and our physical world. This new collectivism is one in which the ecosystems we depend on are members of the collective. Our actions must promote a collective economic allocation of the resources for living in ways and to degrees that do not damage the physical world but are, instead, regenerative. We may call this 'regenerative collectivism', and its central concern is with the well-being of people *and* the planet.

The COVID-19 global pandemic has exposed our collective vulnerability and, as we have discussed in this book, the worst aspects of the environmental crisis are amplifying this. This is fertile ground for the emergence of new forms of collectivism, the foundations of which may already exist. Kallis et al. (2020), for example, identified collectivist responses to the pandemic, particularly in the actions of medical and nursing teams, and they recognise the emergence of a 'care ethic' (p. 9). Although the pandemic has demonstrated collective exposure to public health risks, however, the global experience of the pandemic has not been equal. This is true also of the environmental crisis. As explored in the case study of Vanuatu in Chapter 9, some populations live at the sharp end of immediate climate-related threats. The global rich enjoy greater protection than the global poor in terms of their immediate exposure to negative public health and environmental impacts, but, ultimately, there will be no escape from the impact of major climate-related issues as they accelerate. For us, therefore, our collective futures are likely to be best safeguarded by collective action.

Where, then, might we identify the core value and actions of a regenerative collectivism? Clearly, there are echoes of Enlightenment thought that continues to permeate Western culture and literature, and can, for example, be identified in outcomes as diverse as the US constitution and European liberal and democratic socialism. Although now often rendered obsolete as 'grand narratives' by postmodern social theory, ideas of equality, citizenship, and government obligations to the less well-off remain fundamental to liberal democracy. It is also true that alternative grand narratives, which are increasingly challenging liberal democracy, exist in the form of economic nationalism, xenophobia, and racism. These are becoming increasingly influential in the various populist political movements that have emerged. It is, perhaps, ironic that these threats to liberal democracy appear to be a response to the failure of capitalism to offer hope to significant sections of the global population. In the rust belt towns of the USA and the post-industrial communities of Europe, for example, we see the rise of prejudices that seek to scapegoat 'the other' for the plight of the redundant white working class.

Towards 'regenerative collectivism'

In attempting to contend with climate change mitigation, governments are increasingly being attracted to efforts to develop *technical* solutions to decarbonise the economy. The proposed 'gas-led' recovery from the COVID-19 pandemic in Australia, for example, typifies this approach, along with faith placed in developing the hydrogen economy and carbon capture, both largely unproven technologies (Hepburn, 2020), and the latest IPCC report warns of potential rebound effects from certain technologies (IPCC, 2022). Whilst highly desirable in many respects, renewable energy development also risks underpinning a continued attachment to high consumption patterns and undermining the UN SDG 12. Indeed, the SDGs have failed, thus far, to challenge the globally dominant growth paradigm (Spaiser et al., 2017). As we discussed in Chapter 2, concepts such as de-growth, post-growth, inclusive growth, and green growth remain relatively fringe activities undertaken by think tanks, academics, environmental NGOs, and relatively small groups of engaged citizens. This book contends, that if we are to address the environmental crisis, then continued global growth is impossible. This raises the question, however, of how *after growth* economics is to be achieved, without the potential collapse of employment in global carbon-intensive industries and the mass social disruption that may follow.

For us, a prerequisite to change is the necessity for a 'just' transition (see: Newell and Mulvaney, 2013; McCauley and Heffron, 2018). Such a transition must develop patterns of meaningful work and economic reward, and must ensure globally equitable solutions to achieving carbon reduction. This requires social rather than simply technological innovation. It also requires new forms of economic organisation that prioritise social development, equality of access to health and educational services, and the promotion of universal community well-being. As we have discussed at various points in this book, there are novel economic models that either seek to modify capitalism in order to achieve these objectives or propose more radical departures from the global capitalist system. Some of these have gained considerable traction with municipal authorities, regional governments, and civil society organisations. Despite a multiplicity of proposed alternatives, however, there remains little debate about the mechanics of broader economic change. How does the global transformation of our systems of production and consumption happen?

Some insight into this problem may be gained from a brief revisit to the 'modes of production' debate of the 1970s Althusser and Balibar, 1970). Whilst now arcane and of little general relevance, there are some insights from this debate that may shed light on our current concerns. The 'mode of production' concept relates to the assemblage of social, political, and ideological relationships that give coherence to an economic order. For Marx, these patterns are distinguished in the historical epochs of slavery, feudalism, and capitalism. Each was characterised by the domination of one class by another, in which the product of labour was appropriated by the dominant class. This historical relationship, for Marx at least, would end with the creation of the communist mode of production in which ownership of the means of production by one class would be abolished. In detailed, and at times obscure, debate, these patterns were elaborated further during the 1970s and 1980s to consider whether, rather than dividing history into distinct epochs, there have been periods when multiple modes of production have co-existed (Benton, 1984). From this perspective, although it was generally agreed that each mode of production did, indeed, contain the 'seeds of its own destruction', there might be a period of coexistence between the old and the new mode of production. Without diving too deeply into this arcane debate, the point here is that it is possible for a replacement for capitalism to coexist with capitalism.

The many examples of innovative economic models identified at points throughout this book are experiments to potentially identify a new mode of production to supersede contemporary capitalism or to radically modify it. Rather than a struggle for dominance, however, a multiplicity of innovative models is likely to contribute to an embryonic replacement for current models of capitalism, which no longer exploit people or planet. As we have indicated, such an alternative model may be considered a form of regenerative collectivism, where the purposes of 'production' are human well-being and planetary regeneration.

On transition to a new order

Regenerative collectivism builds on the essential elements of a range of collectivist paradigms drawn from Western philosophical and political ideas, as well as highlighting the value of traditional cultures with patterns of common ownership and environmental stewardship. It points to the need for new thought systems that recognise the environment as a core element of the collective. Regenerative collectivism extends its membership to the biosphere and the myriad of species that inhabit it, of which humans are just one. We need to work with nature, rather than against it. Forms of collectivism have, of course, existed throughout history, but have often been obscured by more vociferous and pernicious ontologies.

The distortion of human history by Social Darwinism, for example, based entirely on the belief in species development and survival through conflict and competition, is itself an ideological outcropping of capitalism. In this crudely deterministic rendering of Darwin's theory of evolution, excessive individualism, dominance by the most able, and the validity of racial hierarchies were echoed in the competitiveness of the capitalist market system (Kallis et al., 2020). This creed of 'survival of the fittest' has been used to justify everything from the natural superiority of the ruling class, to the racial supremacy of white Europeans. This coalesced to justify the exploitation of what it called 'inferior classes' within the emerging capitalist economies and was honed to justify the enslavement and genocide of indigenous people in the process of colonisation that progressed during the 18th and 19th centuries. The class structure and racism of the era served the development of capitalist trade and the accumulation of wealth that became the basis for further inequalities.

The imperialist period also saw the celebration of the 'triumph of man' over the natural world, which became the basis of our collective blindness to environmental degradation. The perspective of the planet as an unlimited resource, and the prospect of eternal economic growth, also have their origins in the colonial era. The world was, quite literally, for our taking. The civilising force of capitalism was presented as a beacon, bringing light into the 'dark continents' and the 'primitive' people who inhabited them. Capitalism was ideologically constructed as a civilising force, bringing freedom to peoples enslaved by tribal cultural and social patterns. The reality of the enslavements, genocides, and economic despoliation of the colonial period was justified by the complex configuration and articulation of evolutionary theory, religion, racial superiority, classical liberalism, and market capitalism.

In contrast, collectivism has often been an intrinsic aspect of human history. Communities of people have often developed collaborative social systems based on kinship and tribal traditions of exchange and reciprocity. These are often balanced with a careful stewardship of nature. More fundamentally, alternative understandings of evolutionary theory suggest that natural selection by cooperation is the central characteristic of all species. Kropotkin (1902) identified the 'mutual aid' that characterises almost all intra-species relationships, even in highly territorial species that cooperate over their boundaries. In carefully documented studies conducted in Siberia, Kropotkin charts the universality of mutual aid. Biologist Lynn Margulis (as cited in Jackson, 2021, pp. 84–85) was keen to point to the fact that animal and plant life evolved through symbiosis – evolution happens as much through cooperation as through competition. Capitalism, in contrast, is premised on individualism and competition:

> Among characteristics unique to the historical complex leading to global growth, individualism is among the most deeply internalized and difficult to address. How to disentangle the roles that individualistic institutions have played in advancing certain rights, freedoms, and equalities from the roles they have played in undermining others, including rights to common management of territories, resources, and cultures for sustaining shared wellbeing?
>
> (Kallis et al., 2020, p. 28)

The struggle between individualism and collectivism has manifested itself in historical competition between the institutions of capitalism and the organisations that have sought to represent labour, which had their origins in the immiseration of working people. Ironically, the success of capitalism is itself the triumph of collective political organisation and the political capture of social and political institutions by elites and their representatives. In modern societies and cultures, collectivism has been a major driver of social change and the improvement of the human condition. In examining the 'golden age' of welfare capitalism in Chapter 3, we saw improved well-being through the, albeit short-lived, burgeoning of the welfare state. Even in austerity riven Britain, the basics of health, education, and social security systems remain weakened, but intact. We make a very clear distinction between the enforced collectivisation of communist regimes and associated ideologies of collectivism and the ethical practice of principles of altruism and mutual aid. The latter are immutable aspects of human behaviour, which wither or flourish depending on the dominant hegemony. The result is contingent on the ideology it is articulated by, and the position it is ascribed in the hegemonic struggle.

The environment as a member of the collective

During the 20th century, expressions of collective existence and manifestations of 'social contract' were codified within complex legal, welfare, and social frameworks, and tended to be encapsulated in the term 'human rights'. Such rights were politically enunciated in the 1948 United Nations Declaration of Human Rights, and since then nations have, to varying degrees, encoded these within constitutions, legislation, and judicial precedent. These 'anthropocentric' rights (Miller, 1998) have consequently become the foundation stones of legal protections in 'developed' nations, as well as conditioning international law. Emerging from this has been the idea of a human right to a sustainable environment. A sustainable environment is, from this perspective, one that is not exhausted by over-exploitation of natural resources or polluted by excessive production and consumption. On a global scale, this requires us to work collaboratively towards positive environmental stewardship and to address the many aspects of the environmental crisis we have discussed in this book. This may be described as 'environmental justice' (Hiskes, 2009).

The concept of environmental justice also extends to those yet unborn future generations, who will inherit the global environment and our impact upon it (Duwell and Bos, 2018). Hiskes (2009) suggested we are connected to future generations through shared values and identities, creating reciprocity between present and future regenerations that extends the rights we enjoy to those who follow us. This future generations' perspective has become a powerful tool in the advocacy for positive environmental stewardship. In Wales, this perspective has led to future generation legislation, although as we discussed in Chapter 2, this has not yet fully influenced wider economic policy. The logical implication of this discussion is that future generations must also form part of the regenerative collective.

Following the extension of anthropocentric rights to the physical world, more recently there has been the emergence of conceptions of the environment as an entity with its own 'ecocentric rights' (Miller, 1998). Miller defines an environmental right as: '…the right to a continued existence unthreatened by human activities…attached to non-human species, to elements of the natural world and to inanimate objects' (Miller, 1998, p. 6). This appears to contradict the multiplicity of beliefs that see the physical world as the domain of 'man' to be utilised as assets. Nevertheless, there has been a long history of affording the natural world a degree of legal protection. In the UK, protection was provided to certain seabirds as early as 1869, and there followed many statutes '…which offer direct protection of landscape and of various species of plants and wild animals' (Miller, 1998, p. 159). Indeed, Miller identified over 109,000 square kilometres of land enjoying a range of statutory protection (p. 161).

Although there is significant historical precedent in the UK and elsewhere of declaring legal protections for land and other species, this has tended to be limited to national parks and reserves, whilst vast deforestation, asset extraction, air, soil, and water pollution, and species extinction have continued elsewhere. The inclusion of the entire environment within the collective could reinvigorate human relationships with nature. This is an important point because it rejects the view that we have 'environmental externalities' and related spatial fixes. The 'collective' implies ecological and social care and restoration as universals, not as opt-ins. Whilst there will no doubt be abuses of ecocentric rights, the wider inclusion of the environment within the collective would more forcefully enable the denunciation of breaches and ensure stronger legal recourse than is currently the case. It is important to emphasise, of course, that the requirement for a cultural 'adoption' of the environment as a member of the collective is also paramount. No amount of legal framing will provide sufficient protection for the environment unless there is a collective inclusion of it as an equal member. Regenerative collectivism must engender a reconnection with nature, and this raises some important questions about extending common rights to lands and the biosphere as we discuss in Chapter 4.

We can again turn to indigenous cultures for some guidance on what might be achieved. Marshall (2019) suggested that the conferment of what she termed 'legal personality' or 'legal personhood' (p. 234) to natural entities such as rivers, contradicts the 'earth community' philosophy of First Nations people, and continues a process of colonisation by removing indigenous land ownership, and the millennia-old obligations of stewardship of the land enshrined in aboriginal laws and cultural practices (p. 236). The creation of land as a legal entity, from Marshall's perspective, actually separates it from humans, whereas in aboriginal culture and language no distinction is made between 'country' – the air, water, and people who inhabit it. Whilst many of Marshall's concerns relate to the legal frameworks and historical dispossession of First Nations people from their land, it should be noted that even by international comparison Australia has very few constitutional and legal protections of the environment (Pepper and Hobbs, 2021). Nevertheless, it is important that the extension of regenerative collectivism to the environment does not accelerate the historical separation and alienation of humans from the environment.

Exploring contemporary regenerative collectivisms

Throughout this book, we have explored various examples of regenerative collectivism, that have sought to establish innovative place-based economic relationships that provide sustenance and welfare to the global populations and avoid the excessive inequities that are inherent in growth-based neoliberalism, and/or seek to address the environmental crisis. We do not propose to restate all these examples here, but it is worth noting just a few of the examples that have been discussed.

Foundational Economy

In Chapter 7, we highlighted the work of the Centre for Research on Social and Cultural Change (CRESC) at Manchester University on what they termed the 'Foundational Economy' (Bentham et al., 2013). As we have noted, interest in the Foundational Economy appears to echo Blumenfeld's (1955) discussion of the basic and non-basic economy, where 'the basic economy' consists of those activities that citizens supply to one another. For CRESC, the problem was that certain economically 'chosen' sectors, those believed to offer the most growth potential, tend to offer little from which to construct stable and sustainable economies and are, in any case, spatially imbalanced. Moreover, as most advanced governments have pursued similar strategies global competition is intense. Consequently, CRESC called for a far greater balance in economic policy, and for a focus on '...welfare-critical goods and services...' (Froud et al, 2018, p. 19). Despite its many virtues, this approach has been criticised for its lack of engagement with the environmental agenda (Sayer, 2019), a criticism that it has more recently sought to address. Despite these limitations, it is difficult to conceive of regenerative collectivism without an economic model that does not include a focus on human well-being.

The Commons

In Chapter 4, we discussed the plight of the Commons. Following the pioneering work of Elinor Ostrom (1990), there has been '...an explosion of literature on commons that developed the concept in relation to diverse goods and resources such as information, open-source software, genetic code, seeds, food, land, housing, urban space, firms and credit' (Lukas, 2021). Ostrom demonstrated that the 'tragedy of the commons' thesis – which argued that commonly held natural resources (e.g. grazing land) would inevitably be destroyed by the self-interested actions of participants – was false. At the core of conceptions of the Commons lies a collectivist ethic, which can be identified in '...collective entrepreneurial experiments, cooperatives, community-based enterprises and peer production initiatives' (Albareda and Sison, 2020, p. 731). Esteves et al. (2021) described the burgeoning Commons economy as the 'social solidarity economy' (SSE) (Esteves et al., 2021, p. 1424). They argued that the 'institutionalisation' of the UN SDGs and their adoption by organisations within the SSE will help expand the Commons. Commons have the potential to avoid issues associated with the private market and state control of key resources, and, therefore, have the potential to significantly add to regenerative collectivism.

Community Wealth Building

In Chapter 7 we referenced approaches to community wealth building, which have emerged to help address the failure of orthodox economic policy to meet the needs of post-industrial communities (discussed in Chapter 3). Often emerging at a local level from community development programmes, community wealth-building initiatives seek to provide business development and employment generation activities that are not dependent on inward investment. Dubb (2016) identified two core strategies associated with this approach: leverage of existing expenditure from place based public and non-profit organisations (the 'anchor institutions' we identified in Chapter 7); and wider anchoring tactics that seek to develop businesses (generally not-for-profit) that can supply goods and services required by anchor institutions. With its origins in many US cities during the 1960s, community wealth building continues to attract place-based experiments to realise its potential. Just as, for example, Wales experimented with the Value Wales Toolkit, other places such as Preston have attempted community wealth-building initiatives (CLES and Preston City Council, 2019). Such initiatives can be constrained by procurement legislation and existing practice, and the failure of community wealth-building approaches to adequately integrate ecological imperatives is a major issue. Notwithstanding these caveats, Community Wealth Building would seem to offer much to regenerative collectivism.

Doughnut Economics

As we discussed in Chapter 2, Kate Raworth's (2017) 'Doughnut Economics' is visualised as a doughnut shape diagram, with its social foundations and ecological ceilings. The model is concerned with human deprivations (such as hunger and illiteracy) and planetary degradation (such as climate change and biodiversity loss). The space between the 'doughnut' represents the safe operating zone for people and planet. Raworth argues that such an honest appraisal prompts us to, for example, change the economic goals, see the big picture (including the creativity of the commons), nurture human nature, design to overcome inequality, create to regenerate (a circular not linear economy) and be agnostic about growth. As we have indicated throughout this book, such thinking is closely aligned with, and indeed influenced, our own.

Regenerative collectivism and place

We experience our social interactions, conflicts, networks, and cultures of human connectedness largely through the context of physical places. It is, of course, all too possible to romanticise the concept of community, whilst ignoring its negatives such as its mechanisms of social control, criminal victimisation, ethnic division, and cultural oppression. Place, however, tends to inform our individual and collective consciousness. The community development strategies associated with the urban regeneration programmes identified in Chapter 3, for example, help demonstrate the deep capacity of communities to foster reciprocity, volunteering, and mutual aid.

This sense of community solidarity is, perhaps, easier to envisage in small towns or rural communities, but it is not by any means limited to them. Urban collectives can create micro-communities within anonymity of the metropolis. Community gardens, food sharing schemes, community libraries, and car-share schemes are just some examples of such collective actions. Dr Joan Clos (2017) (then Executive Director of UN Habitat) referred to the 'street' in urban social life, a collection of meetings, liaisons, and connections that take place in urban spaces, which create a social life grounded in community connections. This is, perhaps, epitomised in the 'superblocks' strategy in his home city of Barcelona. Utilising the unique existing block structure of the Eixample district of the city to combine nine blocks, the strategy limits vehicular access, reduces vehicle speed to a pedestrian friendly 10 kph, and consequently returns the use of the street to the 260,000 residents. The scheme is generally regarded as a very successful environmental and social intervention (López et al., 2020), notwithstanding the risks of gentrification that often accompany urban renewal schemes and require careful management of socio-spatial outcomes (Scudellari et al., 2020). Similar reorganisations of urban space have the potential to reenergise community and social connections to create urban collectives.

Place is also where we experience our most direct connection to the environment. In urban regeneration programmes, primary concerns of residents tend to focus on 'crime and grime' (Adamson, 2010). Grime refers to the environmental degradation of estates and neighbourhoods in which participants experienced the decay of physical amenities. Participation in clean-up activities, urban gardening schemes, and environmental improvements is often encouraged to promote and develop long-term volunteering commitments. In the opening paragraph of this book, we referred to the experience of place as primarily an environmental experience, which shapes our childhood and, ultimately, our life chances. Whilst our connection to the environment can be abstract and based on wider concerns for planetary well-being, it is fundamentally rooted and conceived in the place we live.

Places offer a new nexus for theory and practice

In Part One, we sought to reformulate the concepts and theories that are central to debates about the twin and combined social and environmental crises – the core concerns of this book. We also sought to expose the nature and extent of those crises in light of these new perspectives, and how counter-hegemonic ways of understanding economics, social structures, the environment, and our political cultures have conditioned our current responses and the solutions advocated. In Part Two, we sought to demonstrate in an empirical way how the necessary changes to economic and social praxis we advocate in earlier chapters may be brought about through a place-based methodology, which seeks to empower places and achieve real change. The Deep Place case studies discussed in Chapters 8 and 9 represent our own modest experiments into how the necessary changes might be practicably implemented through a place-based Coalition for Change. This, we suggest, could and should become the locus for future action involving the public, private and third sectors in partnership with local communities.

Within the Coalition for Change model, we have the embryonic elements of regenerative collectivism. As we have seen, to date the response to the various Deep Place studies has been relatively piecemeal, only in some areas has the governance model been more fully embraced. That said, many of the proposals contained within the studies have influenced specific policy responses within communities and, we would suggest, have (along with some of the other complimentary models we have highlighted) begun to permeate wider political, economic, and environmental policy discourse. In Wales, for example, well-being legislation encompasses many of the tenets we have advocated in this book. In 2018, the Senedd created the Welsh Youth Parliament, empowering younger people to get involved in decision-making. The Welsh Government has been experimenting with the Foundational Economy, there has also been a greater emphasis on local public sector procurement and the role of anchor institutions, and, most importantly, Wales has declared nature and environmental emergencies. Much more needs to be done, but the beginnings of change are noticeable.

We believe that ideas and practices formulated at a local level can 'travel up' the policy superstructure to organically influence wider change and establish the kinds of ideas we advocate in this book. That, for us, represents the means by which to establish and reinforce the new politics and economics required for regenerative collectivism to grow and ultimately flourish. The process will not be easy, and the countervailing and reactionary theories outlined in Part One, and existing mainstream thinking, will continue to push back against change. Place, however, is the context within which the roots of regenerative collectivism may be firmly established. Places can elucidate pathways through the social and environmental crises that have been the focus of this book. Places are the location of micro-activities that induce macro levels of change. *Places offer hope.*

References

Adamson D (2010) *The Impact of Devolution: Area-Based Regeneration Policies in the UK.* York: JRF.

Albareda L and Sison A J G (2020) 'Commons organizing: embedding common good and institutions for collective action. insights from ethics and economics', *Journal of Business Ethics*, 166(4): 727–743. https://doi.org/ 10.1007/s10551-020-04580-8.

Althusser L and Balibar E (1970) *Reading Capital*. London: New Left Books.

Bentham J, Bowman A, de la Cuesta M, Engelen E, Ertürk I, Folkman P, Froud J, Johal S, Law J, Leaver A, Moran M and Williams K (2013) *Manifesto for the Foundational Economy*. Manchester: Centre for research on Socio-Cultural Change, University of Manchester.

Benton T (1984) *The Rise and Fall of Structural Marxism. Althusser and His Influence*. London: MacMillan.

Blumenfeld H (1955) 'The Economic Base of the Metropolis: Critical remarks on the "basic - nonbasic" concept', *Journal of the American Institute of Planners*, 21(4): 114–132. https://doi.org/10.1080/01944365508979342

CLES (Centre for Local Economic Strategies) and Preston City Council (2019) How we built community wealth in Preston. Achievements and lesson. Manchester: CLES. Accessed 7 February 2022: https://www.preston.gov.uk/media/1792/How-we-built-community-wealth-in-Preston/pdf/CLES_Preston_Document_WEB_AW.pdf?

Clos, J (2017) 'Opening conference address'. *Implementing the New Urban Agenda in Australia and the Pacific*. International conference, Melbourne. 4 May.

Dubb S (2016) *Conversations on Community Wealth Building*. Maryland: Democracy Collaborative.

Duwell M and Bos G (2018) 'Why Rights of Future People', pp. 9–27 in: M Duwell, G Bos and N van Steenbergen (eds.) *Towards the Ethics of a Green Future: The Theory and Practice of Human Rights for Future People*. London and New York: Routledge.

Esteves A M, Genus A, Henfrey T, Penha-Lopes G and East M (2021) 'Sustainable entrepreneurship and the Sustainable Development Goals: community-led initiatives, the social solidarity economy and commons ecologies', *Business Strategy and the Environment*, 30(3): 1423–1435. https://doi.org/10.1002/bse.2706.

Froud J, Johal S, Moran M, Salento A and Williams K (2018) *Foundational Economy*. Manchester: Manchester University Press.

Fukuyama F (1992) *The End of History and the Last Man*. New York: Free Press.

Hepburn S (2020) '4 reasons why a gas-led recovery is a terrible and naïve idea', The Conversation. Accessed 21 January 2022: https://theconversation.com/4-reasons-why-a-gas-led-economic-recovery-is-a-terrible-na-ve-idea-145009.

Hiskes R (2009) *The Human Right to a Green Future. Environmental Rights and Intergenerational Justice*. New York: Cambridge University Press.

IPCC (2022) Climate Change 2022: Impacts, Adaption and Vulnerability. Summary for Policymakers. Accessed 1 March 2022: https://report.ipcc.ch/ar6wg2/pdf/IPCC_AR6_WGII_SummaryForPolicymakers.pdf

Jackson T (2021) *Post Growth: Life after Capitalism*. Cambridge: Polity.

Kallis G, Paulson S, D'Lisa G and Demaria F (2020) *The Case for DEGROWTH*. London: Polity.

Kropotkin P (1902) Mutual Aid: a factor of evolution. Accessed 7 February 2022: https://www.marxists.org/reference/archive/kropotkin-peter/1902/mutual-aid/index.htm.

López I, Jordi O and Mercedes P (2020) 'Mobility infrastructures in cities and climate change: an analysis through the superblocks in Barcelona', *Atmosphere*, 11(4): 410.

Lukas P (2021) *Democracy, Markets and the Commons: Towards a Reconciliation of Freedom and Ecology*. Bielefeld: Verlag.

Marshall V (2019) 'Removing the Veil from the "Rights of Nature": the dichotomy between first nations customary rights and environmental legal personhood', *Australian Feminist Law Journal*, 45(2): 233–248. https://doi.org/10.1080/13200968.2019.1802154.

McCauley D and Heffron R (2018) 'Just transition: integrating climate, energy and environmental justice', *Energy Policy*, 119(April): 1–7. https://doi.org/10.1016/j.enpol.2018.04.014.

Miller C (1998) *Environmental Rights. Critical Perspectives*. London and New York: Routledge.

Newell P and Mulvaney D (2013) 'The political economy of the "just transition"', *Geographical Journal*, 179(2): 132–140. https://doi.org/10.1111/geoj.12008.

Ostrom E (1990) *Governing the Commons. The Evolution of Institutions for Collective Action.* Cambridge: Cambridge University Press.

Pepper R and Hobbs H (2021) 'The Environment Is All Rights: Human Rights, Constitutional Rights and Environmental Rights', *Melbourne University Law Review*, 44 (2): 634–678. Accessed 7 February 2022: https://search.ebscohost.com/login.aspx?direct=true&AuthType=sso&db=lpb&AN=151532395&site=eds-live&scope=site&custid=s2775460.

Raworth K (2017) *Doughnut Economics: Seven Ways to Think Like a 21st-Century Economist.* London: Random House Business Books.

Sayer A (2019) 'Moral economy, foundational economy and decarbonisation', *Renewal: A Journal of Social Democracy*, 27(2): 40–46.

Scudellari J, Staricco L and Brovarone E V (2020) 'Implementing the Supermanzana approach in Barcelona. Critical issues at local and urban level', *Journal of Urban Design*, 25(6): 675–696.

Spaiser V, Ranganathan S, Swain R and Sumpter D J T (2017) 'The sustainable development oxymoron: quantifying and modelling the incompatibility of sustainable development goals', *International Journal of Sustainable Development and World Ecology*, 24(6): 457–470. https://doi.org/10.1080/13504509.2016.1235624.

INDEX

Note: **Bold** page numbers refer to tables; *italic* page numbers refer to figures and page numbers followed by "n" denote endnotes.

Printed in the United States
by Baker & Taylor Publisher Services

Printed in the United States
by Baker & Taylor Publisher Services